I0462708

2010

MAJOR REVISION
The best gets even better!

With 2011 supplements!

End User License Agreement

This book (the "Book") is a product provided by MechboxPRO.com o/b AirsoftPRESS (being referred to as "MechboxPRO" in this document), subject to your compliance with the terms and conditions set forth below. PLEASE READ THIS DOCUMENT CAREFULLY BEFORE ACCESSING OR USING THE BOOK. BY ACCESSING OR USING THE BOOK, YOU AGREE TO BE BOUND BY THE TERMS AND CONDITIONS SET FORTH BELOW. IF YOU DO NOT WISH TO BE BOUND BY THESE TERMS AND CONDITIONS, YOU MAY NOT ACCESS OR USE THE BOOK. MECHBOXPRO.COM MAY MODIFY THIS AGREEMENT AT ANY TIME, AND SUCH MODIFICATIONS SHALL BE EFFECTIVE IMMEDIATELY UPON POSTING OF THE MODIFIED AGREEMENT ON THE CORPORATE SITE OF MECHBOXPRO.COM. YOU AGREE TO REVIEW THE AGREEMENT PERIODICALLY TO BE AWARE OF SUCH MODIFICATIONS AND YOUR CONTINUED ACCESS OR USE OF THE BOOK SHALL BE DEEMED YOUR CONCLUSIVE ACCEPTANCE OF THE MODIFIED AGREEMENT.

Restrictions on Alteration

You may not modify the Book or create any derivative work of the Book or its accompanying documentation. Derivative works include but are not limited to translations.

Restrictions on Copying

You may not copy any part of the Book unless formal written authorization is obtained from us.

LIMITATION OF LIABILITY

MechboxPRO will not be held liable for any advice or suggestions given in this book. If the reader wants to follow a suggestion, it is at his or her own discretion. Suggestions are only offered to help.

IN NO EVENT WILL MECHBOXPRO BE LIABLE FOR (I) ANY INCIDENTAL, CONSEQUENTIAL, OR INDIRECT DAMAGES (INCLUDING, BUT NOT LIMITED TO, DAMAGES FOR LOSS OF PROFITS, BUSINESS INTERRUPTION, LOSS OF PROGRAMS OR INFORMATION, AND THE LIKE) ARISING OUT OF THE USE OF OR INABILITY TO USE THE BOOK. EVEN IF MECHBOXPRO OR ITS AUTHORIZED REPRESENTATIVES HAVE BEEN ADVISED OF THE POSSIBILITY OF SUCH DAMAGES, OR (II) ANY CLAIM ATTRIBUTABLE TO ERRORS, OMISSIONS, OR OTHER INACCURACIES IN THE BOOK.

You agree to indemnify, defend and hold harmless MechboxPRO, its officers, directors, employees, agents, licensors, suppliers and any third party information providers to the Book from and against all losses, expenses, damages and costs, including reasonable attorneys' fees, resulting from any violation of this Agreement (including negligent or wrongful conduct) by you or any other person using the Book.

Miscellaneous.

This Agreement shall all be governed and construed in accordance with the laws of Hong Kong applicable to agreements made and to be performed in Hong Kong. You agree that any legal action or proceeding between MechboxPRO and you for any purpose concerning this Agreement or the parties' obligations hereunder shall be brought exclusively in a court of competent jurisdiction sitting in Hong Kong.

Preface

Airsoft has been well established in Asia for over 20 years, but only in recent years has there been much interest in the Western world. Most professional literature on Airsoft were written in Japanese, with very few translated works available. Even though we are seeing more and more tech tips (in English) popping up on the internet these days, too many of them were written by newcomers who don't really know what they are talking about. If you upgrade your AEG based on their advices, you may risk running into unexpected troubles.

At MechboxPRO (a subsidiary of AirsoftPRESS), we produce technical information based on input from practicing engineers, technicians and field operators who have been with Airsoft since the era of S.S. 9000. Because we are part of the industry, we know what information is really needed, and we make sure our books tell what people really need to know. We do not mind to criticize thing that doesn't work, and we will not hesitate to give you hacks and workarounds to difficult problems. Reading this book should be like having an airsoft professional by your side, passing on useful hints whenever you get stuck.

About this Ultra FPS & ROF book

This Ultra FPS & ROF series book is an extension to the Advanced Mechbox Guide marketed under the AirsoftPRESS brand name. With feedback from customers and reviewers we further develop the original mechbox guide to additionally cover manufacturer specific tech info for FPS/ROF optimization.

To achieve ultra high FPS, you need to make sure the entire mechbox mechanism is optimized in such a way that the desired FPS can be produced without sacrificing ROF and battery life. High FPS performance must be achieved with practicality in mind – you should not need to trade off power with other performance factors, and the gears should not get striped too easily. At the same time, you do NOT want to break the law by shooting over the set power limit.

On the other hand, to achieve ultra high ROF you'll need to make sure the entire mechbox mechanism is optimized in such a way that piston movement is fast enough to produce the rate of fire desired. That means you need to use techniques to reduce the friction introduced due to fast movement, to ensure the spring can rebound fast enough after compression, and to reduce the overall workload of the gears and the motor.

Are you ready for the challenge? ☺

Are you ready for the upgrade works?

Upgrade works can cost you $$$$$. Some upgrade works can also put you into legal troubles. Therefore, before doing anything on the mechbox you better first assess your readiness on several key issues. Start by considering the issue on legitimacy. Is it legal to buy and own airsoft in your area? For teenagers: are your parents going to chew you up for an "unapproved" airsoft purchase?

Legitimacy	
Do they allow airsoft at all?	This is the most important question to ask if your sole source of income is your parents.
Is it legal to own and play airsoft in your area?	Some regions do not allow the purchase of airsoft. Some regions do not allow the possession of airsoft at all. Will you be sent to jail due to airsofting? The best thing to do is to call the local police department and ask. For your peace of mind, ask for a written confirmation from a police officer.
Power limit	If airsoft is allowed, what is the power limit? Check your local laws to make sure you are doing legal upgrades.
Restrictions on full auto shooting	Some countries do not alow full auto airsoft. New Zealand is an example. Some impose power restriction specifically on AEGs. Again, check your local laws.

NOTE: <u>Case study: New Zealand</u>

Air-powered weapons (airsoft guns are air powered) are legal to possess and use in New Zealand, provided that the person is either over 18 years of age, or 16 with a firearms license. A person under 18 may not possess an air gun but may use one under the direct supervision of someone over 18 or a firearms license holder.

It is illegal to use these weapons in any manner that may endanger or intimidate members of the public (pointing, brandishing, etc) except where there is reasonable cause, such as an Airsoft game.

<Police, New Zealand, Airguns Factsheet, http://www.police.govt.nz/service/firearms/info sheet04.html >

The next question to ask is, what are you going to use your airsoft gun for?

"Application"	
What are you going to use the AEG for?	If it is for self-defense or for security (i.e. as an alternative to using real steel), you want a

	seriously upgraded gun that is reliable. If it is for real serious target practicing, you want a longer rifle which can deliver the sort of accuracy you need. If it is for backyard fun only then anything will do ☺ Long VERSUS Short rifle would make a strategic difference in your upgrade goal. Longer rifles place focus primarily on FPS and accuracy, while shorter rifles may opt for higher ROF. We will talk about this later.
If you are going to skirmish frequently with your friends, what do your friends own?	If all your friends use springer only, there is no point of owning a serious AEG. If you guys are forming a team, ask yourself what role you are going to take. A sniper does not use a shorty after all!

 NOTE: When **MAJOR** FPS improvement is the focus, you may want to use 0.25g BBs. 0.20g BBs are way too light at this power level and will not deliver the accuracy you need. If you plan to go over 450FPS+, consider the use of 0.30g BBs for maximum flight path stability. Marushin has some very nice 0.30g BBs (USD$18 something per bag of 1800... ☹).

By the way, heavy BBs do drag down FPS quite a bit (due to the heavier weight) but may produce higher impact (also due to the heavier weight).

> They are VERY expensive though... you should really take this cost element into consideration when defining your desired upgrade level.

Closely linked to the "application" issue is budget. How much $ are you going to spend? Remember, if an AEG costs you $100, you should have at least $300 handy since you will need to buy upgrade parts and tools. You will also need spare $ to deal with parts breakage and all other sorts of expenses.

Budget	
How much $ do you have for the initial investment?	You need enough $ to buy an AEG, a set of battery and charger, and loads of ammos. And you need to have enough $ to get them shipped to your dwelling.
Do you have regular cash inflow?	This is about TCO (Total Cost of Ownership). Some AEGs require more maintenance than the others. You need $ to buy parts, and you need $ to buy tools. Upgraded guns are more powerful and are always more troublesome maintenance-wise! To illustrate: *Our recommendation for maintenance of UPGRADED AEGs (at a minimum) is that for every 20000~30000 rounds, replace the following items:* ● *the hopup rubber and the hopup spacer*

(bucking)

- the motor brushes
- the spring
- the piston head (the O ring in particular)
- the bushings(if they have cracks or they no longer sit on the shell firmly).
- the air nozzle (if its tip starts to deform)

Our recommendation for stock AEGs (at a minimum) is that for every 30000 rounds, replace the following items even if they aren't broken:

- the spring
- the piston head (the O ring in particular)
- the bushings(do this primarily on poor quality metal bushings or plastic bushings)

* also replace the motor brushes and the hopup rubber after about 40000~50000 rounds.

Upgrade Goals

In the European Continent, the most common power limit for AEG is 1J (328 FPS). Certain countries even require that the airsoft replica be restricted to single-shot mode only. There are laws which impose different power limits on different shooting modes, thus rendering full auto shooting almost impractical.

When dealing with laws and regulations, you always want to maintain a safety margin. Remember, there is nothing absolute in the world of airsoft – performance may vary shot by shot. To avoid unintended jail time, our advice is that you set your upgrade/tuning goal to at the max 95% of the legal limit. Say, if the legal limit is 1J, don't go over 0.95J. The small difference in power would not have any practical difference in the war game field anyway.

MechboxPRO AirsoftPRESS (Hong Kong).

Upgrade Strategies

Proper air seal and smooth running of the internals are most critical. Once you have these properly handled, even a stock TM spring can get you to 0.9J or so.

Note the following:

- If what you have is a genuine TM, air seal is good out of the box unless there is heavy wear and tear.

- If what you have is a China-made clone, air seal is terrible but the spring is way over powered. These clones are using a M120 spring to deliver M100 power.

- If what you have is, say, a CA or an ICS, air seal is usually reasonable but there is still room for optimization.

Proper air seal allows for good performance while smooth internals deliver efficiency and reliability. These are the technical requirements that you must meet if proper optimization is to be achieved.

The proper ways of taking things apart and putting them back together

To avoid getting into chaotic situation when doing your assembly and disassembly works, try to be as organized as possible. Do your work on a clean, dry and flat surface which is close enough to reach without having to walk back and forth to access the necessary tools.

In fact, it is best to work things out on a big white towel. The towel provides a color contrast, thus making it easy to see the parts as you lay them down.

Before you remove each part, ask yourself the following questions and take notes if possible:

1. What is this, and what is it for?
2. Why is it made the way it is?
3. How tightly is it screwed on out-of-the-box? How hard is it to remove?

As you remove each part, lay it down on a clean flat surface in clockwise order, with each part pointing in the direction it laid when it was in place. Assign each part a number indicating the order in which it was removed. When you are ready to put them back together, start with the last part you removed and then go counterclockwise through the rest of the parts. If possible, go to your manufacturer's website (or any other web site) and look for any documentation files (mostly in PDF format) they offer for free download. Many of them include very detailed fly-out diagrams, a complete list of all parts as well as where they fit. This will greatly lessen any confusion you might have when putting things back together.

Warning: **!** Along the process of disassembling / assembling your mechbox, there are chances for unexpected debris to fly out (e.g. when you try to put a M120 into the mechbox, your spring guide may accidentally fly out with strong impact). Therefore:

- order your children to stay away from the work area.
- don't work in a location too close to the windows (you don't wanna break the window glass).
- wear safety goggles yourself if you are going to deal with very stiff springs.

Good luck!

MechboxPRO AirsoftPRESS (Hong Kong).

Getting the tools you're going to need

The two basic types of screwdrivers are standard (slot / flat head) screwdrivers and Philips screwdrivers. Make sure you have drivers of different size handy – a whole set of drivers from your local store should cost less than USD$5 each. Only TM mechboxes use Torx screws. Torx head size is typically described using the capital letter "T" followed by a number, such as T5, T10, T15 and T25. TM mechboxes use T10 screws. You therefore need to have the corresponding driver handy.

Note: It doesn't hurt for you to use the regular Philips screws in place of the Torx screws. The Torx screws have nothing special nor unique other than the special screw head layout.

NOTE:  **Is it OK to use flat head screw driver to drive the mechbox torx screws?**

Why not? In fact, as long as the size fits, a flat head driver will work just fine on the torx screws. Do make sure you don't use brute force when driving the screws.

Screws with a centre hole that is hexagonal require the use of Allen wrenches or Hex drivers. To the best of our knowledge, however, no TM mechbox shall require the use of hex screw.

> **You do need to perform motor positioning via a hex screw most of the time.*

Proper handling of the mechbox screws

Most amateurs tend to screw things very tightly in hopes that the parts will not fly off later. This is in fact a deadly mistake because some screws, bolts and nuts are NOT supposed to be tightened too securely or the threads would be stripped.

If you find yourself confronted with a screw that is extremely difficult to get unscrewed, don't use brute force (or you risk stripping the threads). Instead, give the screw a slight twist in the opposite direction and then loose it again. If this does not help, tap the screwdriver on the head with a small hammer (but don't tap it too hard). If it still fails to make it, try to squirt the screw with penetrating oil or WD-40 and retry.

Stripping screws that go into places like the connection points between the mechbox and the pistol grip can be frustrating. If unfortunately you strip a screw, the 2 easy ways to remove it are:

- Fill the stripped screw hole with J.B. Weld (which is a type of glue specially for use with metal parts) or similar kind of product, and then put your screwdriver into the old hole to create a new fitting. Give it 10 to15 minutes to set and dry completely, then unscrew it.

- Drill a hole in the screw, then scoop out the old screw.

On the other hand, to prevent certain critical screws from getting loose, apply threadlock/locktite - a glue type of compound that makes screws more secure. Loctite usually won't bite into plastic very well. It can sometimes soften the plastic, but most of the time it won't really be permanently stuck on there. Most of the time you can get a locked screw loose via the use of a decent screwdriver (by the way, heat is what is used to release excessively strong locktite).

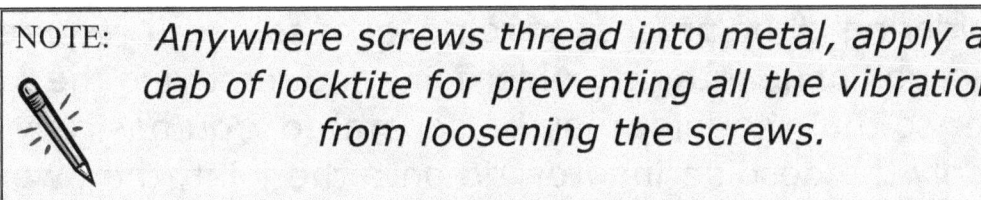

> NOTE: *Anywhere screws thread into metal, apply a dab of locktite for preventing all the vibration from loosening the screws.*

There are locktites of different strength available. You may want to check out the proper type to use through this URL: http://www.loctiteproducts.com/glue.asp

Other tools
You shall need needle-nosed pliers when handling the smaller mechbox screws and springs.

MechboxPRO AirsoftPRESS (Hong Kong).

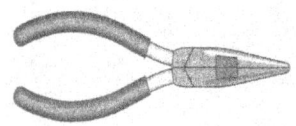

You may need some assorted files (such as straight edge, rounded and rounded side) for deburring and cleaning the edges of cuts needed for slight modification of certain mechbox parts.

You may need to use a pen shape soldering iron for rewiring the mechbox and the battery connection (especially when you are not happy with the existing wiring or you want to switch to the Deans plugs).

Desoldering is the process of removal of solder. It requires the application of heat to the solder joint and removing the molten solder so that the joint may be separated. You basically need to apply the desoldering iron tip onto the joint, then wait for the solder to liquefy, and finally have the molten solder removed through the pump.

If you want to use 7mm or larger ball bearings on standard TM shells, you must enlarge the space holders a bit through an electric dremel tool (those hand held type electric drill like tools for home use would just be fine).

Soldering techniques

The goal of soldering is to join electrical parts together for forming an electrical connection. This is done via the use of a molten mixture of lead and tin (solder) with a soldering iron. Basic supplies needed for proper soldering include a soldering iron (the prong of metal that heats to a specific temperature through electricity), the soldering wire (an alloy of aluminum and lead), and a cleaning resin called flux that ensures the joining pieces are incredibly clean (by removing all the oxides on the surface of the metal that would interfere with the molecular bonding, allowing the solder to flow into the joint smoothly). A perfectly soldered joint should be nice and shiny looking, and should be very reliable in service. The key factors affecting the quality of the joint primarily include cleanliness, temperature, and adequate solder coverage.

The first step in soldering is cleaning the surfaces (including the iron tip itself). They must all be clean and free from contamination. Then, you may melt flux onto the parts to be joined. The parts should both be heated above the melting point of the solder but below their own melting point with the soldering iron. When touched to the joint, this precise heating can cause the solder to flow to the place of highest temperature and makes a chemical bond. Do keep in mind that too much solder is an unnecessary waste but too little may be insufficient for supporting the component properly.

You will find solder paste very useful along the soldering process:

MechboxPRO AirsoftPRESS (Hong Kong).

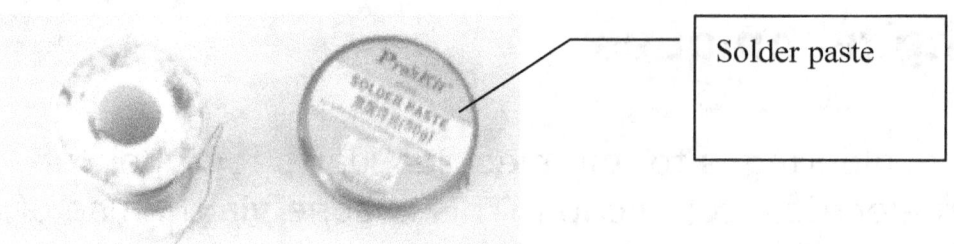

Solder paste

A solder paste is comprised of powdered metal solder suspended in a thick medium, which is known as flux. This flux is added to act as a temporary adhesive for holding the components together until the soldering process is finished.

Solder melts at around 190 degrees Centigrade. Such a temperature is hot enough to inflict a nasty burn. Be extremely careful when you do your soldering work. Also, do your soldering work in a room with good air circulation. Soldering does release toxic fumes.

If you are soldering battery connections, you may want to use a soldering pen of a high WATT value. The one shown below says 60W, which is powerful enough for most soldering jobs. Based on our experience, 20W is way too weak and 40W is only marginally sufficient. A minimum of 60W is recommended.

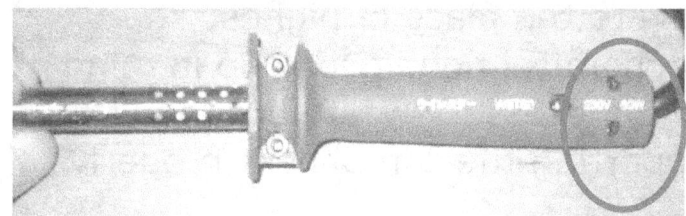

Step by Step Mechbox Upgrade

These pictures show exactly how an AEG mechbox is disassembled for inspection, maintenance and optimization. A V2 based G3 SAS is used for demonstration here.

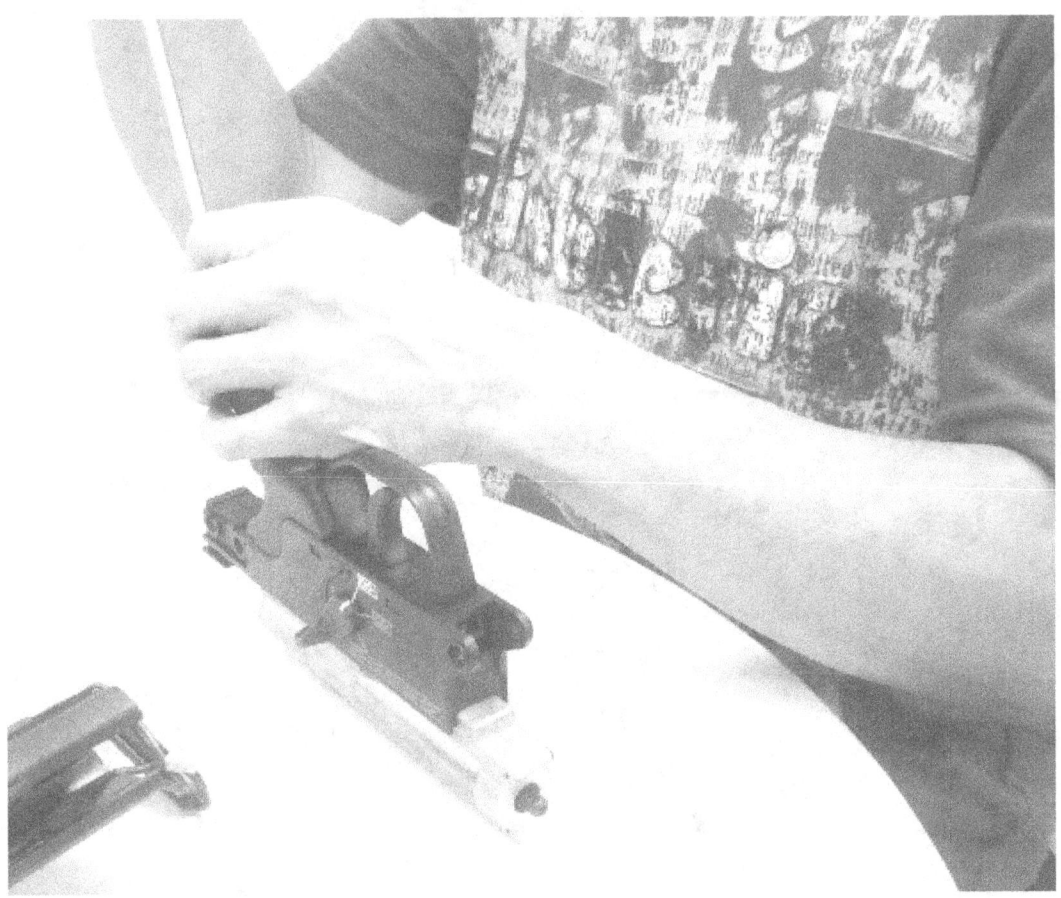

Whenever possible, remove the motor first.

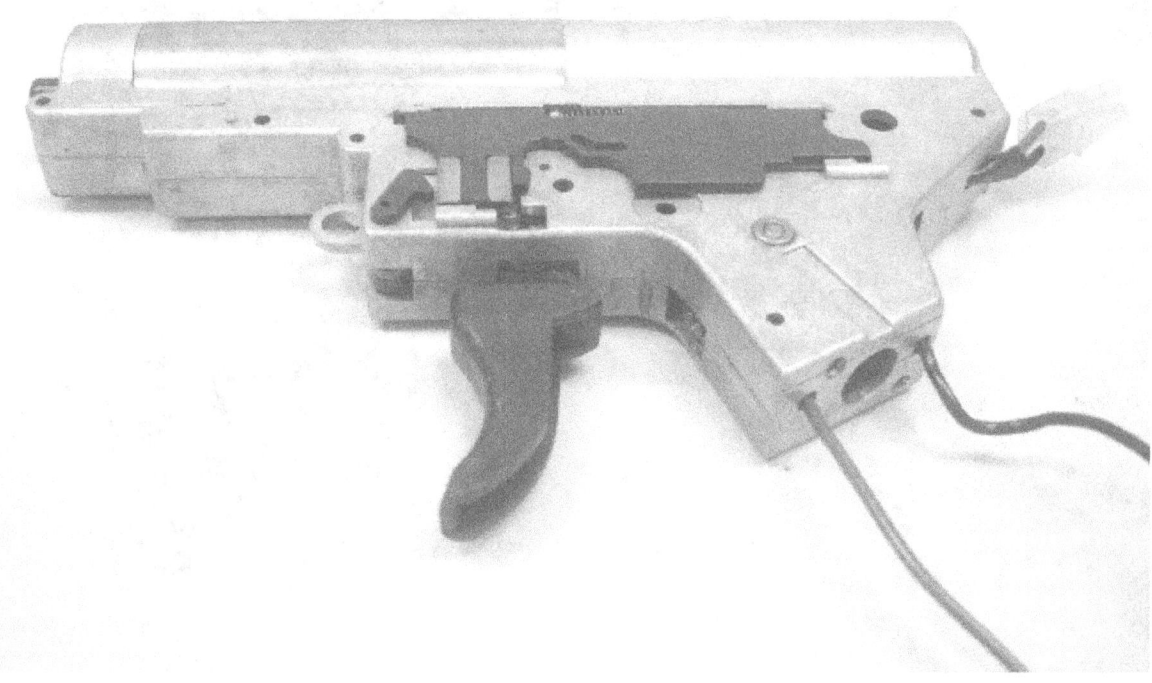

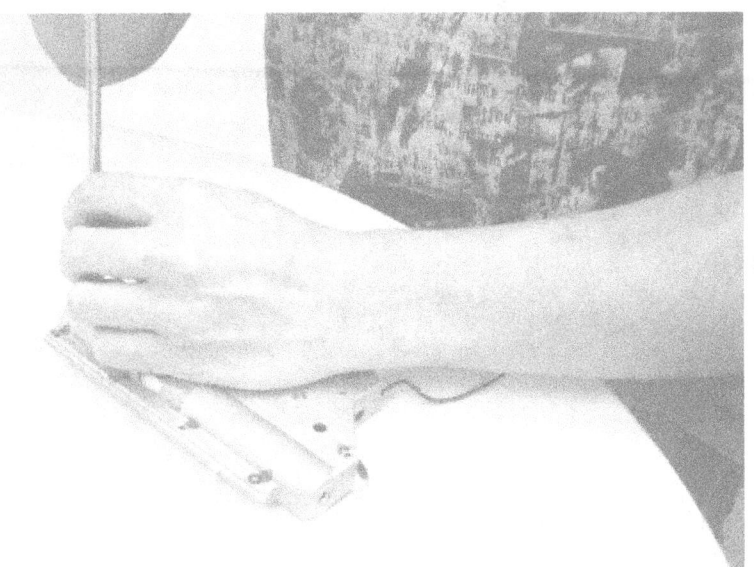

If there are Torx screws, remove them first. In fact, TM uses Torx screws for guiding the order of screw insertion and removal (even though I can hardly see any practical difference).

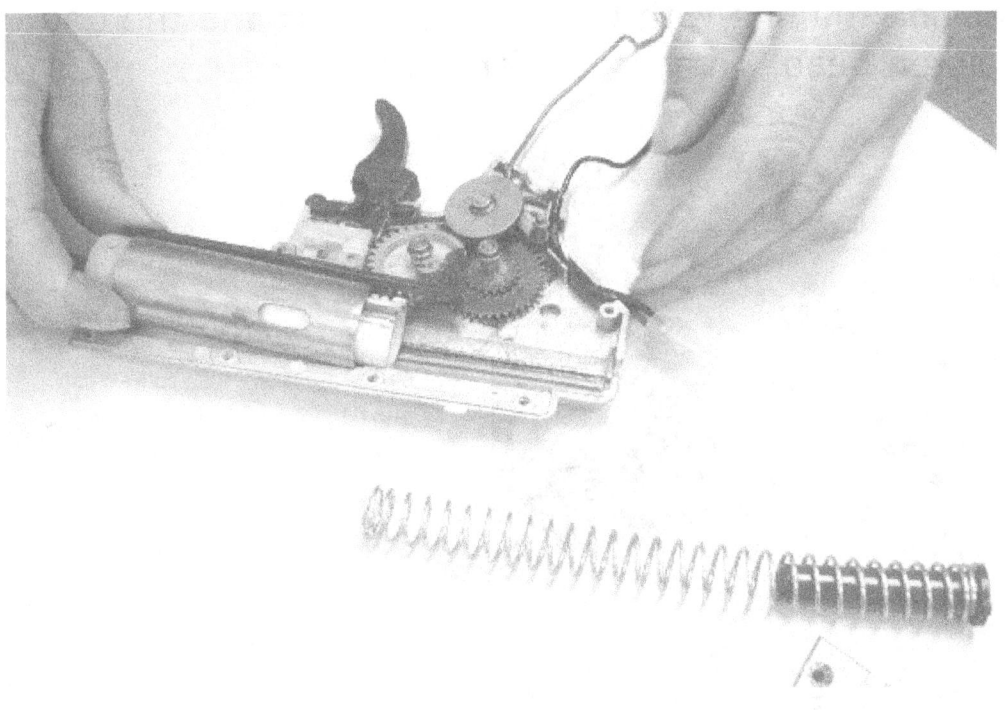

Maintaining Proper Air Seal

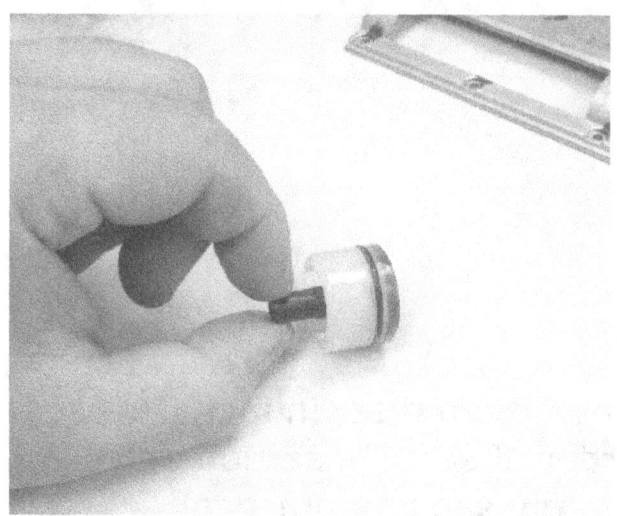

All rubber O rings must be carefully inspected. The two major O rings include the one with the cylinder head, and the one with the piston head. They are differently sized.

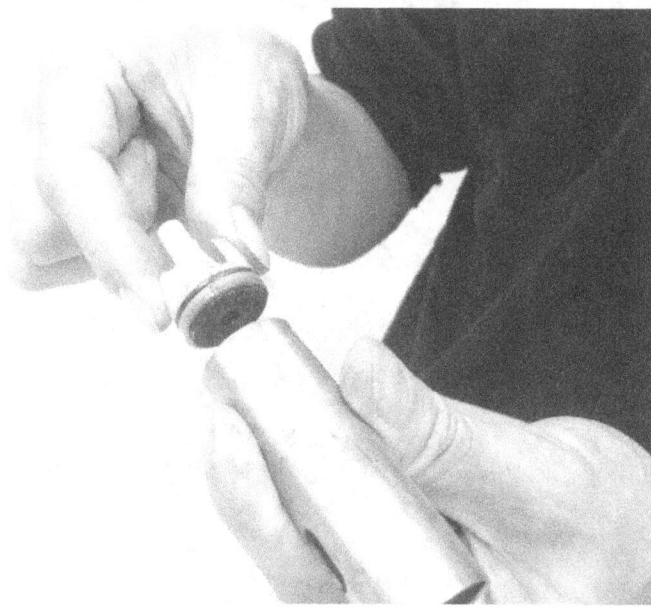

One major problem is that the grease in use actually causes the O rings to "age out" much faster – this explains why the China-made clones are leaky – the grease is bad so by the time the guns reach the consumers the O rings are already damaged.

The cylinder head O ring is made of very thin rubber and is quite difficult to acquire from the local hardware stores (due to its unique sizing). If you don't have one handy, the best thing to do is to apply grease around it for better air sealing. Since the cylinder head does not move, in theory its O ring should not wear out by itself. Still, you may want to regularly check and reapply grease around it if necessary during your regular maintenance effort.

A cylinder nozzle that wobbles could indicate some serious quality problems that you should be aware of... TM cylinders are usually free of this kind of problem but the China-made cylinders are not.

I always prefer a one-piece design here. Like the DEEPFIRE one, it is way more solid. Installation wise it is easy and trouble free.

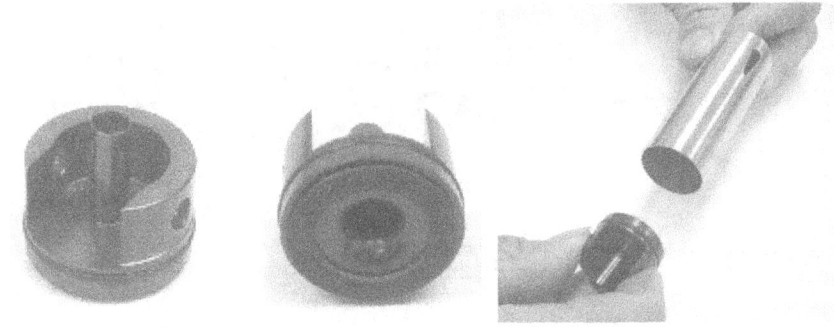

The traditional multi-pieces design looks like this:

Another possible problem is that the inside of the cylinder is not smooth so the piston head O ring is scratched.

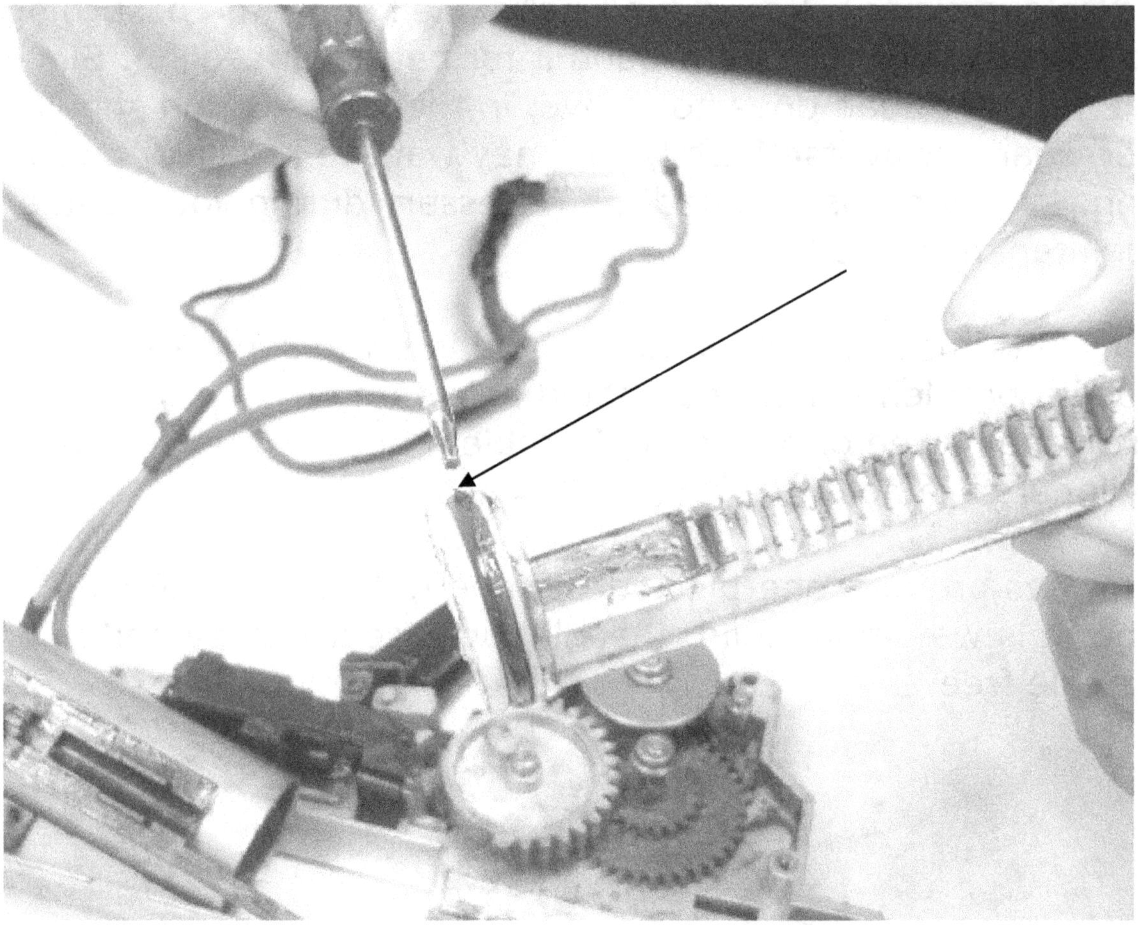

With your finger inside you can feel the smoothness of the inside surface.

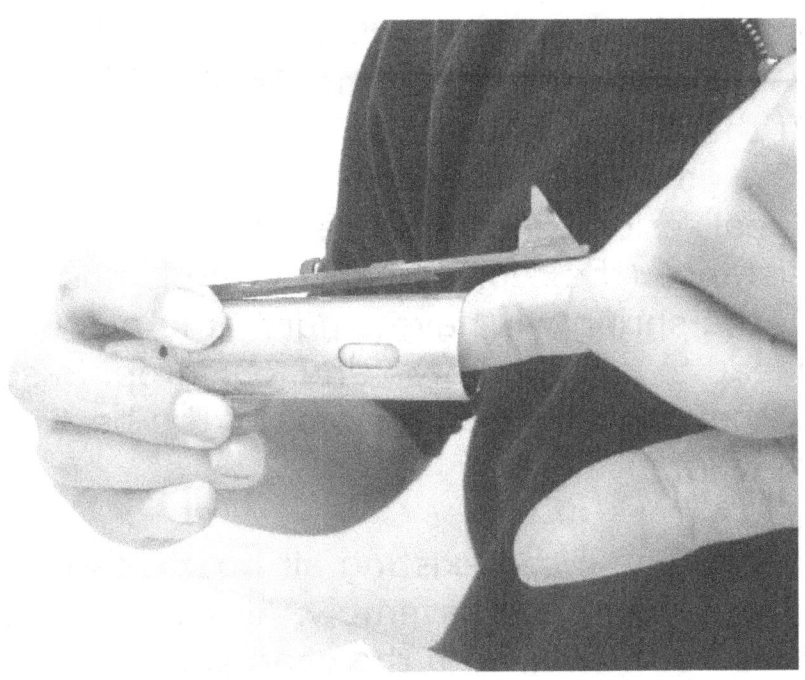

We always suggest that you remove all the existing grease and re-apply good quality grease on all the O rings. The silicon grease must be rubber friendly. And absolutely NO WD-40!

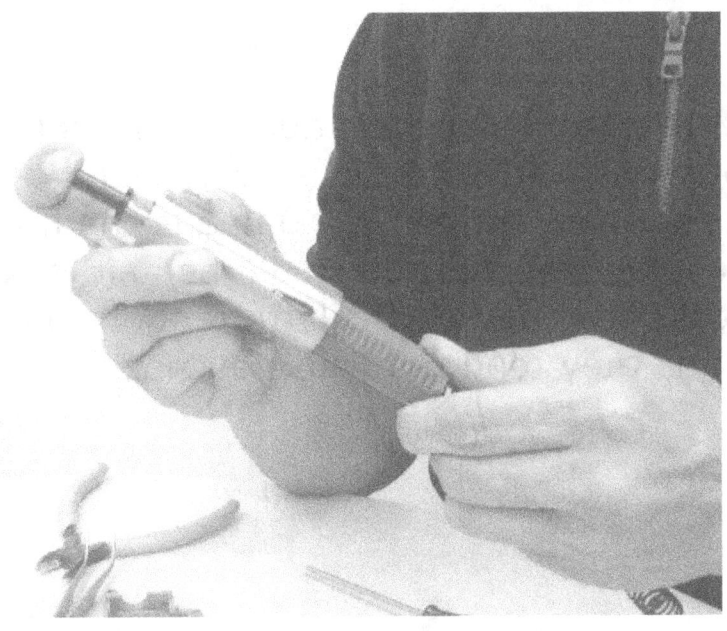

You can test air seal by blocking the nozzle opening and trying to push in the piston. Strong resistance means air seal is reasonably okay. If there is no resistance at all, the piston or the piston O ring is no good (it is allowing too much air leakage, which can result in real serious performance drop). HOWEVER, if there is total resistance (the replacement O ring is possibly too dry or too big), the spring will have a hard time pushing the piston and will lead to both FPS drop and ROF drop. Based on our experience, the stock TM piston is one of the best in terms of smooth movement here.

You then need to determine if the existing air nozzle can produce a good fit with the hopup assembly. Air can still leak out if the nozzle does not fit well with the chamber.

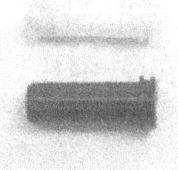

GENERALLY, air nozzle is model specific due to different chamber implementations. If the nozzle is oversized it can hit or scratch the inside of the chamber and damage the hopup rubber. If the nozzle is too short or too small then air leak can result. Do your test fitting – pay particular attention to the China-made clones, as they tend to misuse air nozzle to the fullest extent.

There should be no gap in between. At the same time, the nozzle should not be pushed backward during the test fitting process – if it happens that means the nozzle is too long.

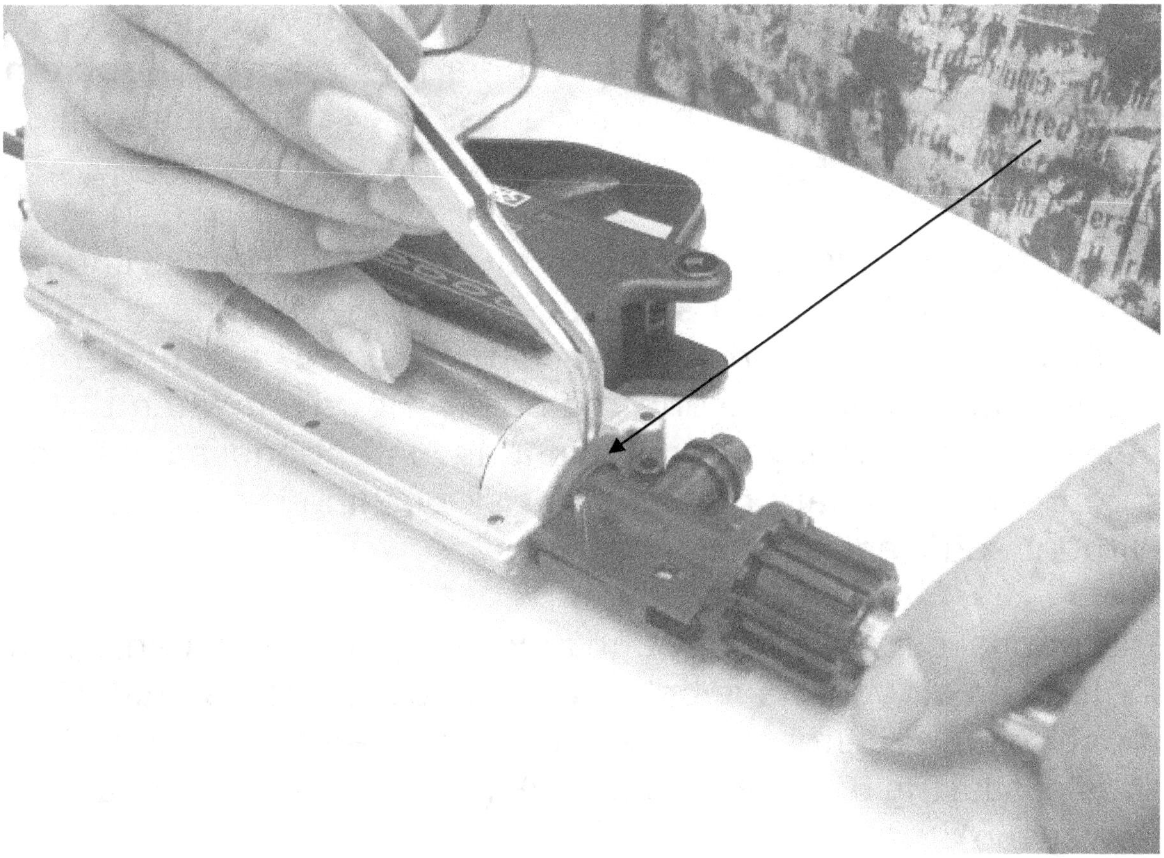

Proper Air Volume Balancing

It is all about barrel length. No problem here with name branded AEGs. Frankly, barrel/cylinder mismatch is a problem only found on the China-made clones. They like to use full volume cylinder on every gun they produce.

When the barrel is too short, too much air from the cylinder will actually affect bullet stability. This is why you need an auxiliary port equipped cylinder so that "extra" air can be leaked out without affecting the bullet. However, when the barrel is a long one, you don't want any intentional air leakage through the auxiliary port or you risk pushing the bullet without sufficient air.

Intentional air leakage through auxiliary port isn't a bad thing. As long as air volume is balanced, the piston can actually move forward faster with the presence of an auxiliary port, thus producing better FPS and ROF (and can conserve battery power too).

 In most cases a larger cylinder is NOT really physically larger. It is all about the effective capacity. The left one is said to be "smaller" because of the opening – the effective capacity starts when the opening "got passed'.

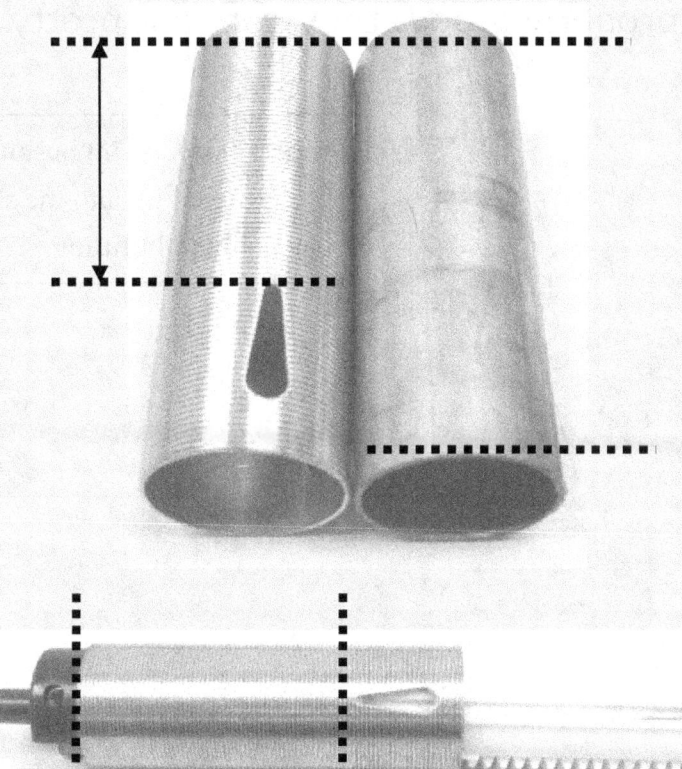

Larger doesn't mean more efficient or effective. For regular barrel lengths (such as those of M4 and MP5) the one shown above with a port closer to the rear side is good enough.

Gears Checking

First of all you need to determine if bushing replacement is necessary. Stock TM plastic bushings are made of good quality nylon and can tolerate inaccuracy in shimming. As long as they are properly lubed, they can last pretty long.

Properly sized 6mm bushings would need to be hammered in onto the shell for a reasonably tight fitting.

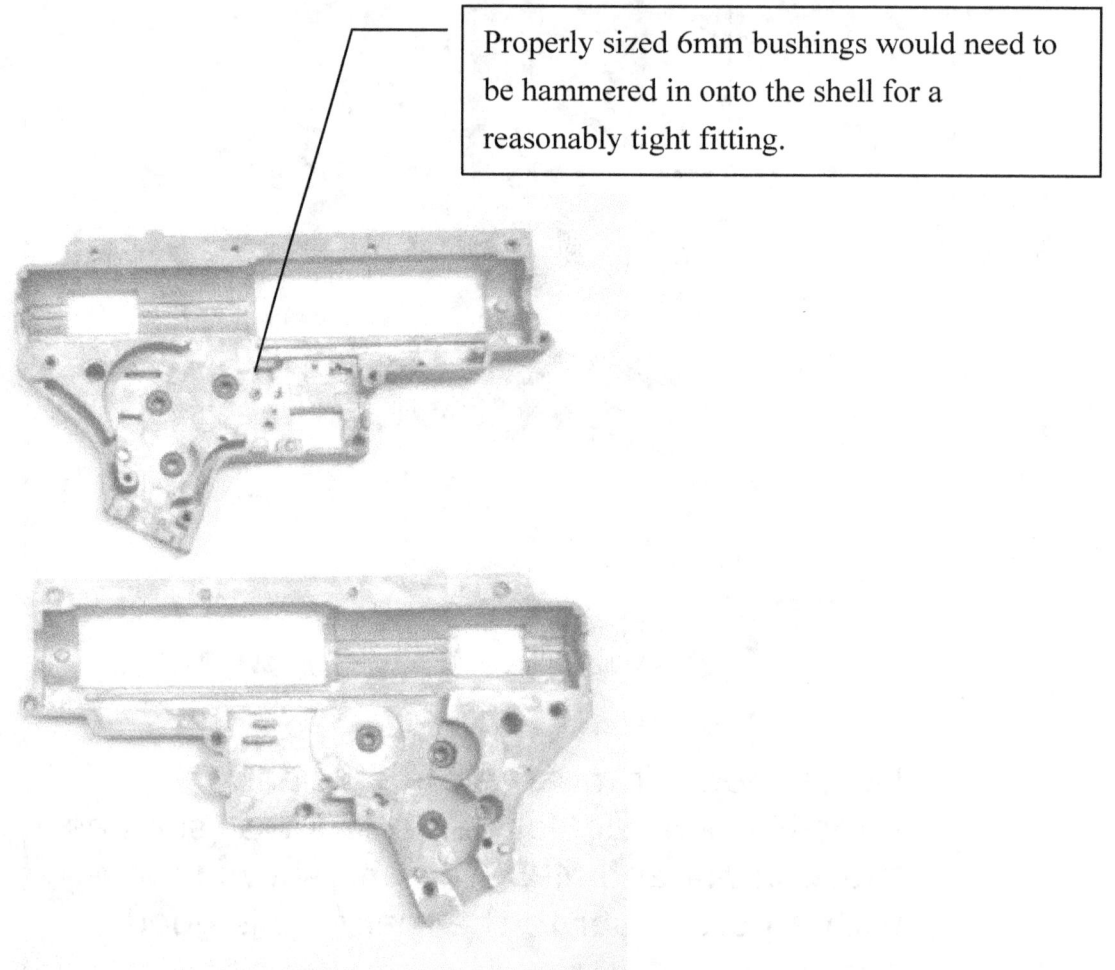

Metal bushings are stronger but they have little tolerance on improper shimming. **In any case, moving to 7mm is unnecessary when dealing with 1J power.**

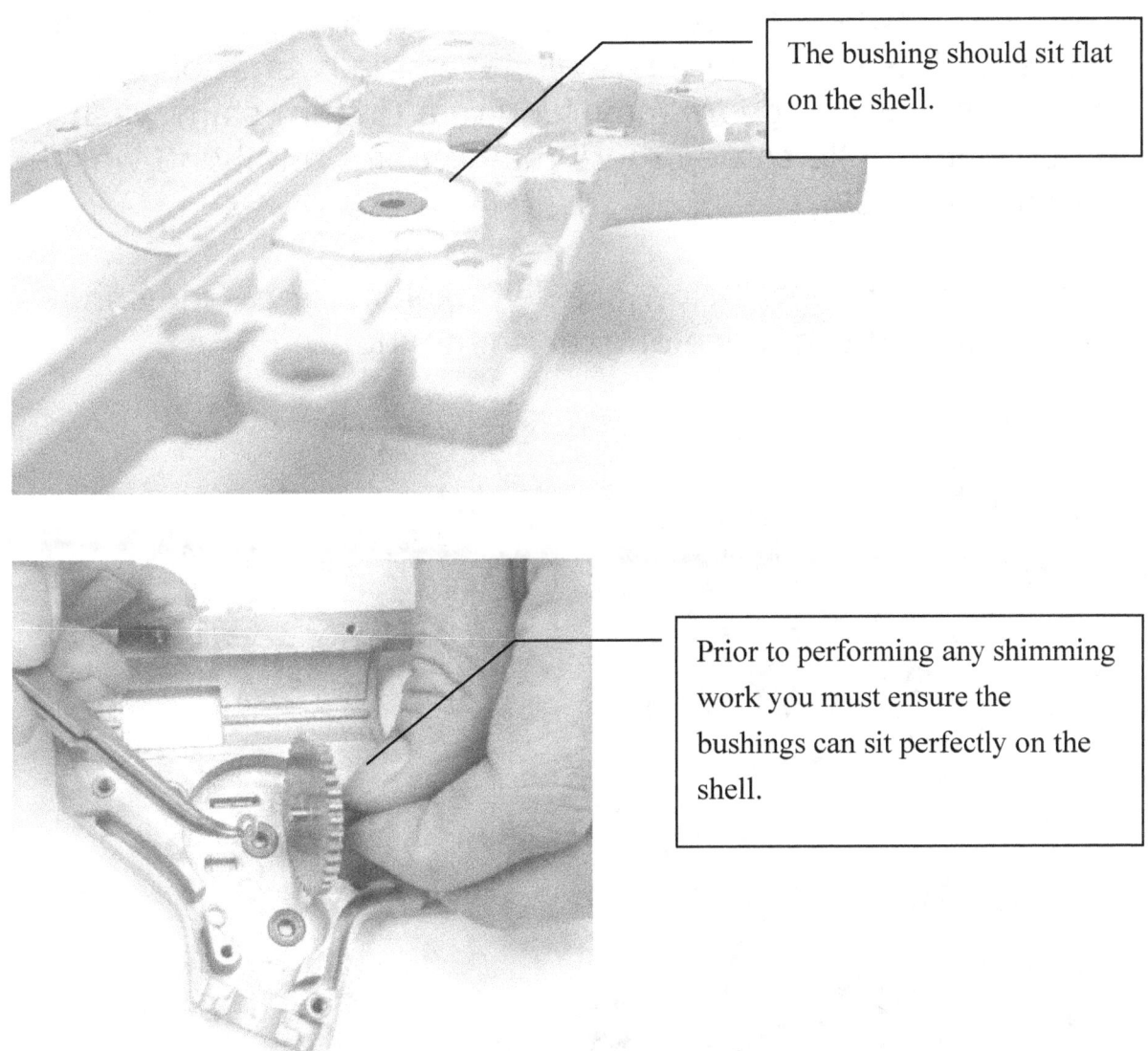

The bushing should sit flat on the shell.

Prior to performing any shimming work you must ensure the bushings can sit perfectly on the shell.

Re-shim whenever you replace the bushings. Start by removing everything on the mechbox shell except for the bushings. By removing almost everything on the mechbox shell you can focus solely on the insertion and shimming of the 3 gears. An empty shell gives you a clearer view of the gears in action.

The reason why you need to carefully shim is that different makes of bushings and gears all have small variations in "thickness". Some gears may require a 0.2mm shim washer on one side while some others may require totally no washer on the top at all.

The series of photos below are self explanatory – they show the suggested sequence of gear shimming:

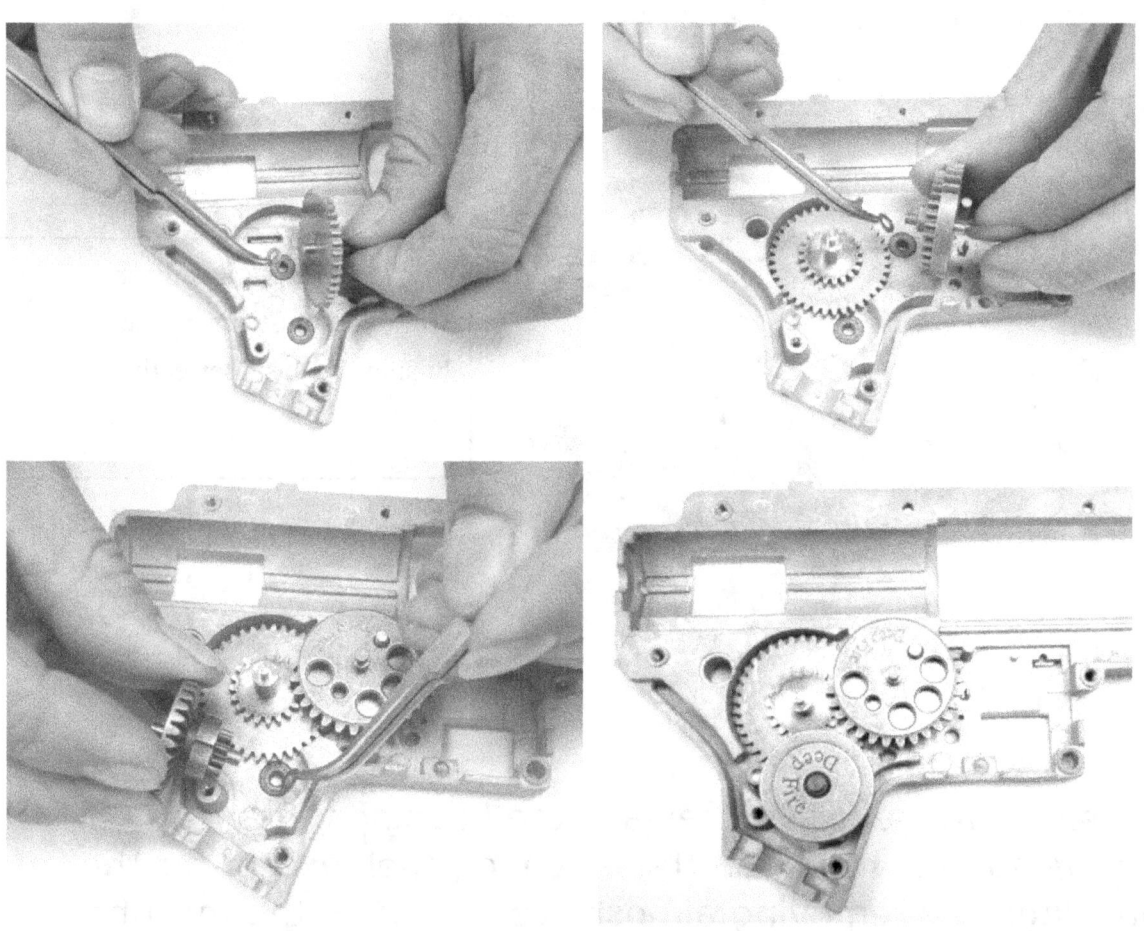

MechboxPRO AirsoftPRESS (Hong Kong).

Look at the setting from the side at different angles (as shown below). This way you can tell whether the teeth of the gears are meshing with the others properly.

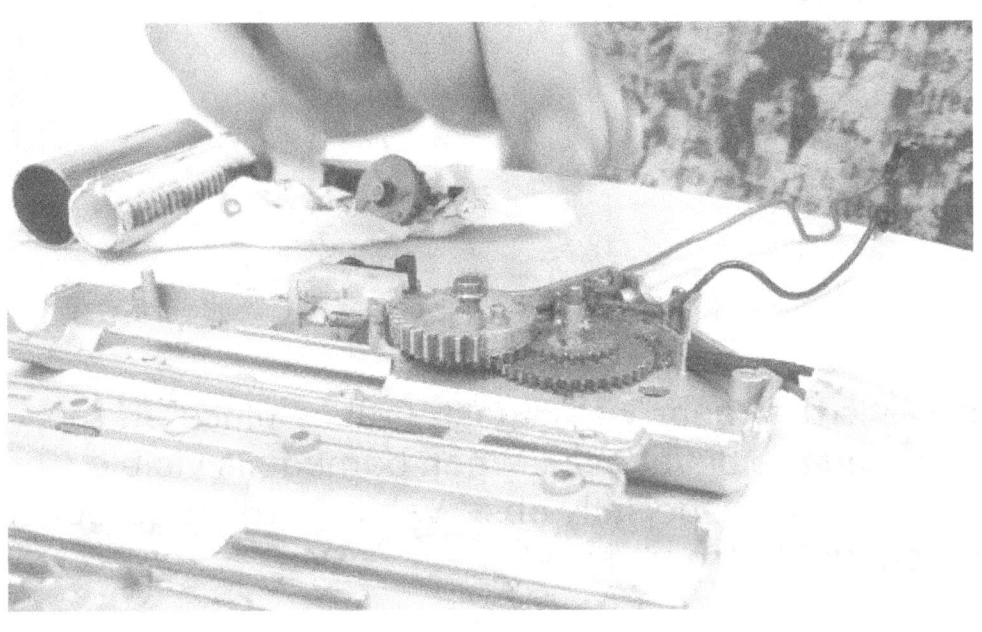

Our experience is that gear wear mostly takes place on the sector gear and on the bevel gear. When damage is visible (for sector gear this can be especially observable on the first tooth), replacement should be considered.

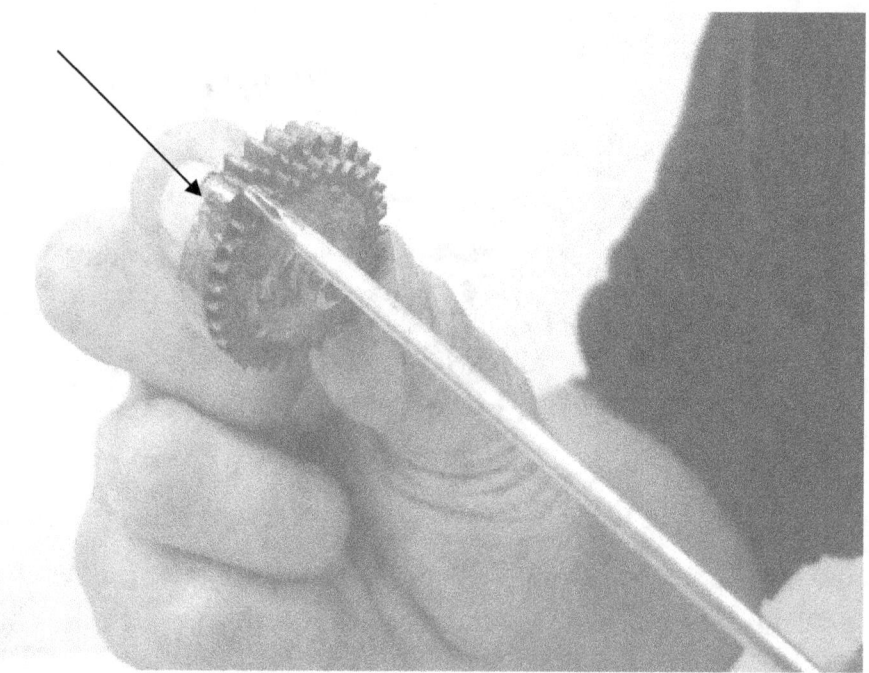

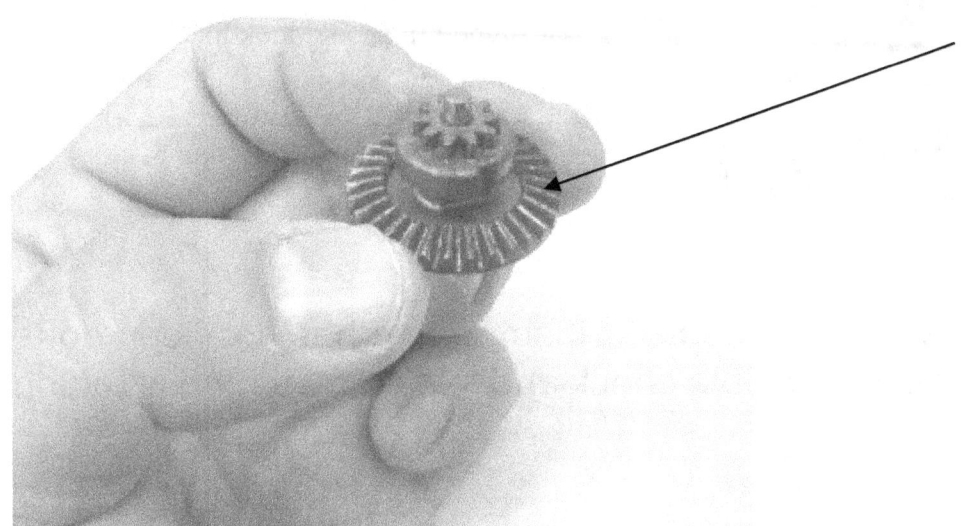

When you see damage on the bevel gear (on the teeth that are in direct contact with the motor), chance is that the motor pinion is damaged as well.

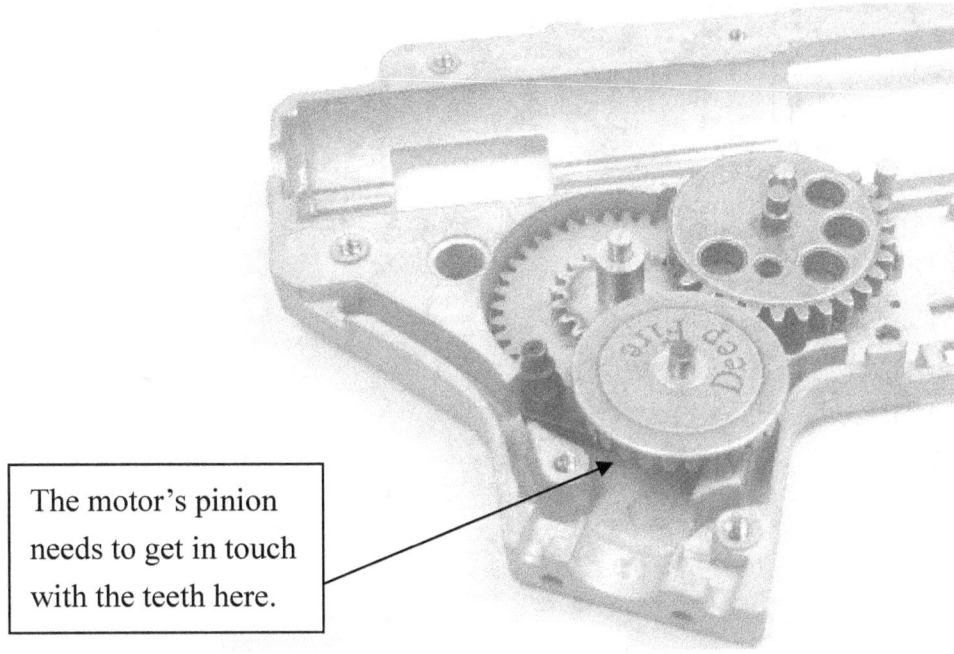

The motor's pinion needs to get in touch with the teeth here.

MechboxPRO AirsoftPRESS (Hong Kong).

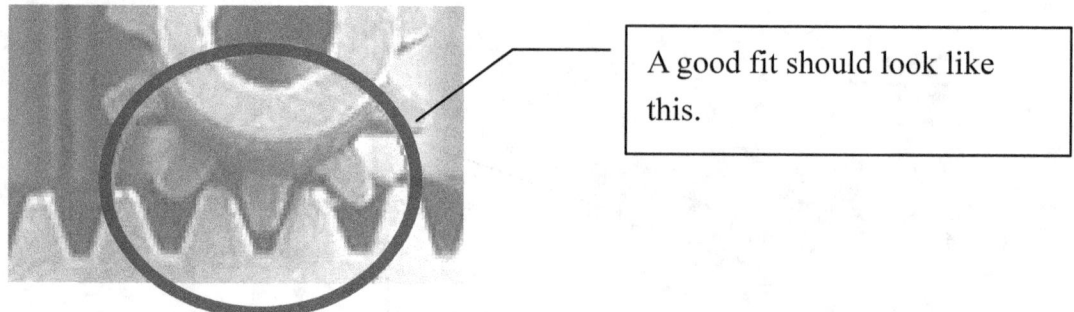

A good fit should look like this.

The anti-reversal can wear out if shimming has not been done properly. Perform a visible check and see if scratching is taking place.

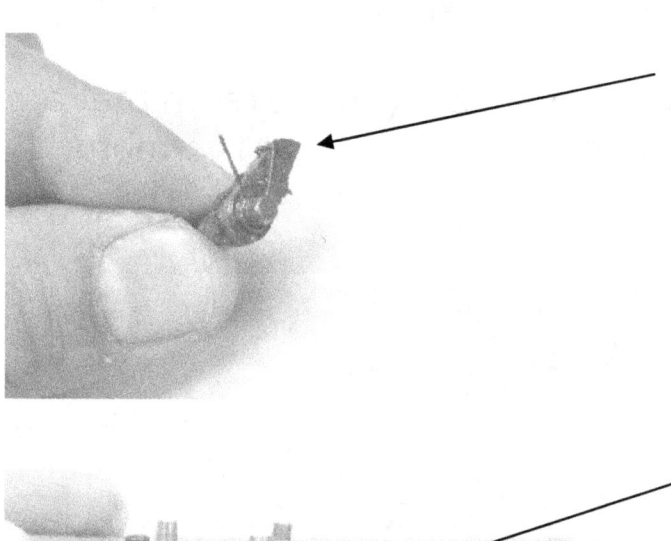

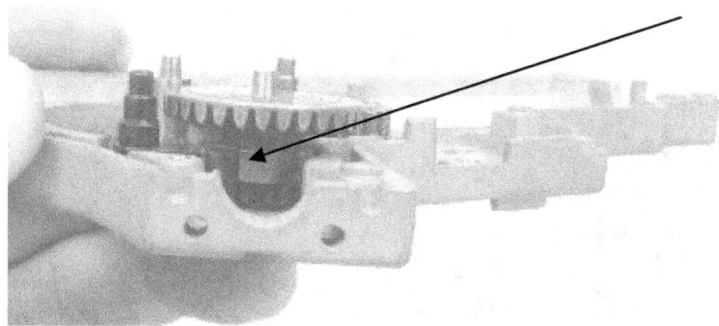

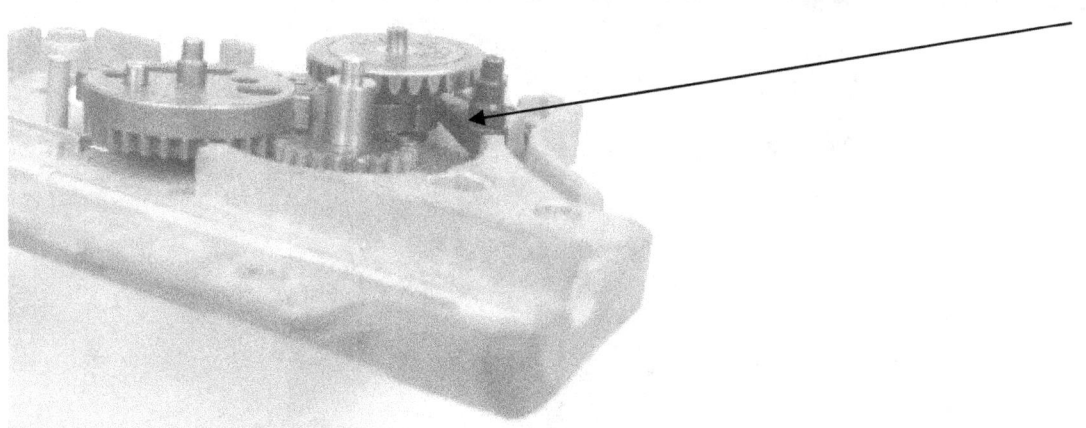

Make sure the anti reversal
would not scratch on the gear.

MechboxPRO AirsoftPRESS (Hong Kong).

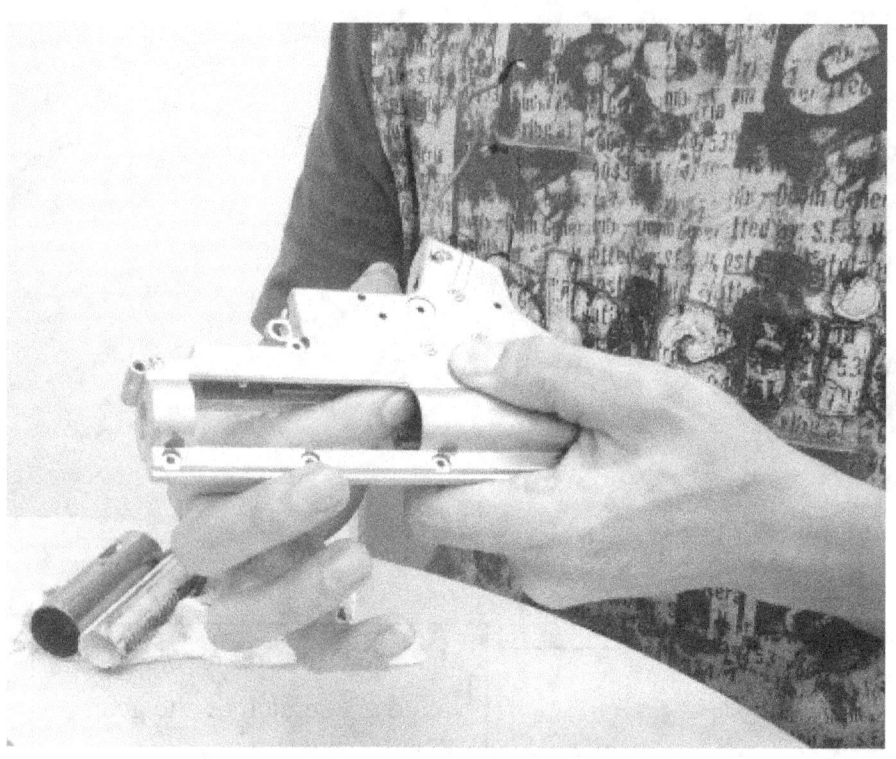

When shimming has been initially completed you need to determine if any of the gears have been over-shimmed. Remember, when the gears run for a long time heat can make them "inflated". Therefore, at the time of shimming you need to leave room for accommodating such "inflation". Tight shimming can produce unexpected "runtime" problems.

If you have difficulties putting the two halves of the mechbox back together, chance is that you have put too many shims somewhere. One extra needless shim on either side of any one gear will usually make it impossible for the two halves to perfectly close.

Once shimming is done you may proceed to configuring the

rest of the internals.

Spring Modification

A stock TM spring is legally safe. A M90 spring is also quite safe. A M100 spring can break the 1J limit quite easily. The stock China-made spring will break the limit for sure.

You may find it necessary to perform spring cut. See the comparison below, a TM spring VS a longer M100 spring. The longer spring needs to be shortened. We prefer to cut the side facing the spring guide for effective power downgrade.

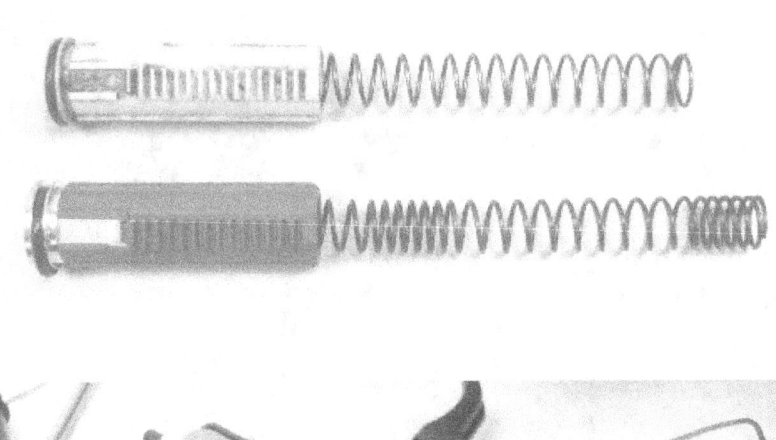

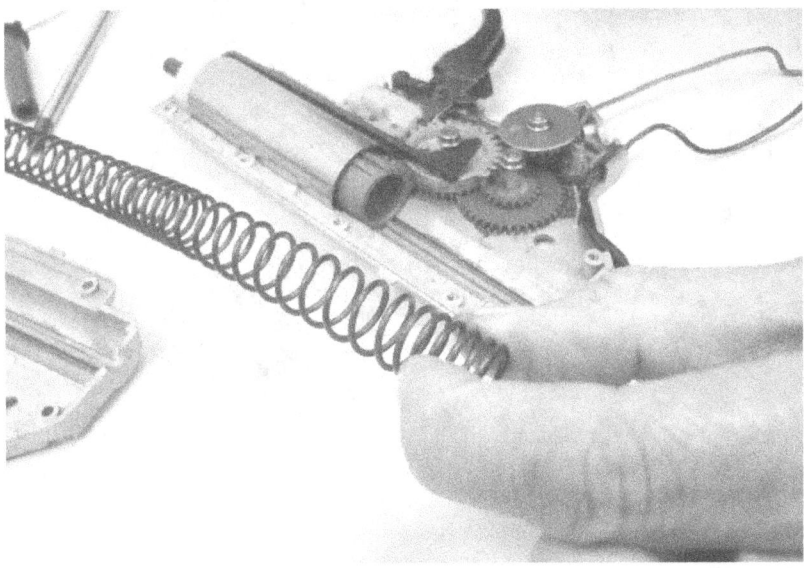

These are the tools you need. First you do the cutting, then you heat up the tail and ground its end to produce a flat & smooth surface (so it can spin freely on the spring guide).

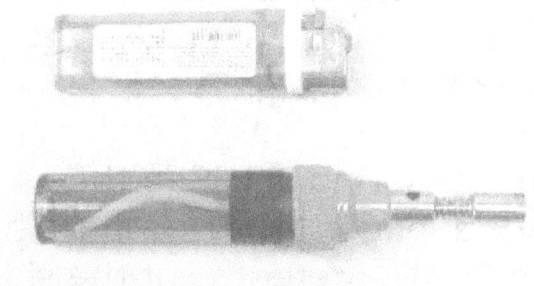

Cutting:

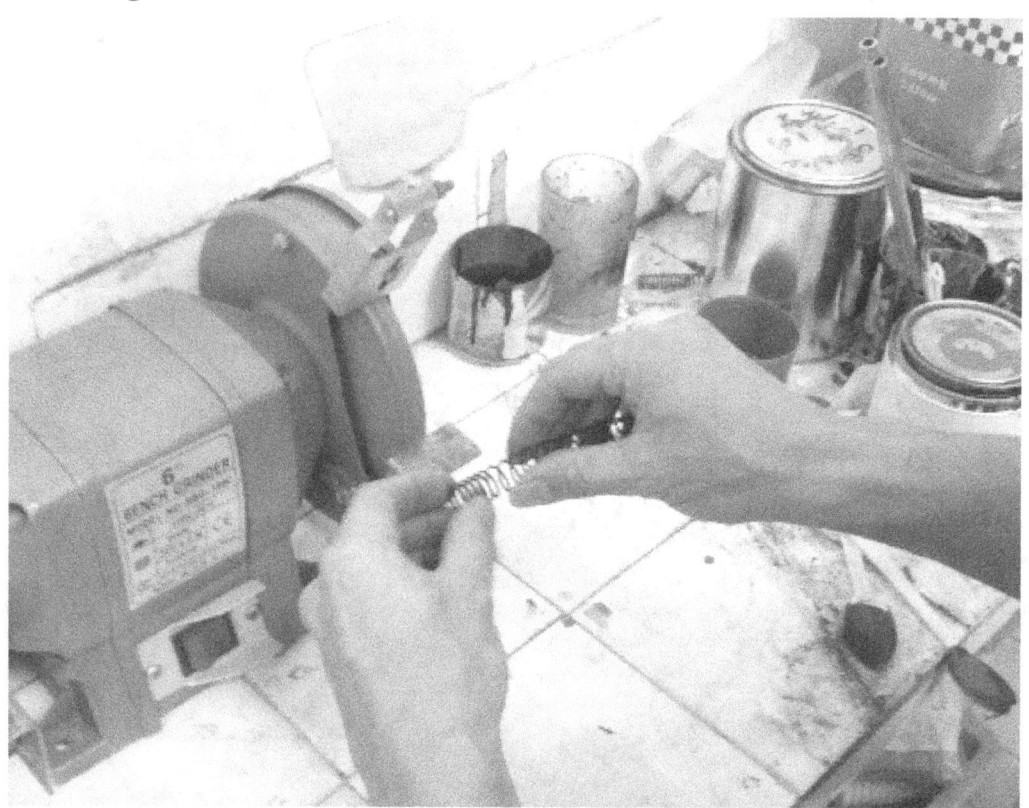

　　　MechboxPRO AirsoftPRESS (Hong Kong).

Heating:

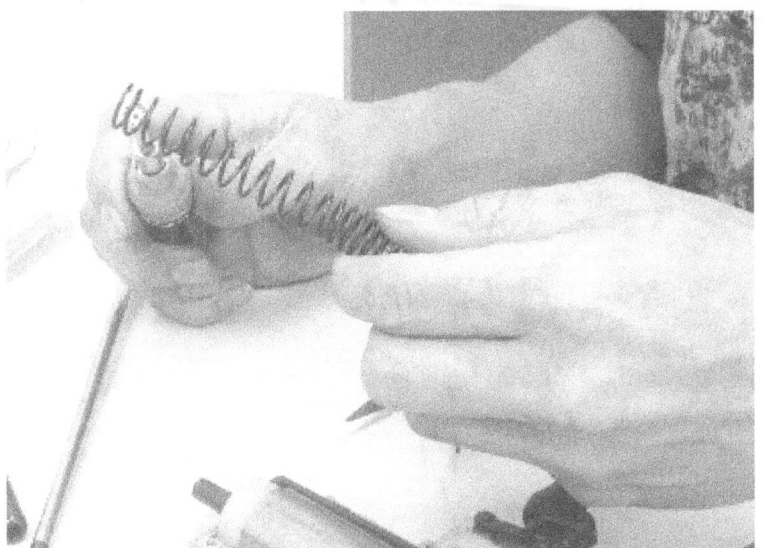

Grounding:

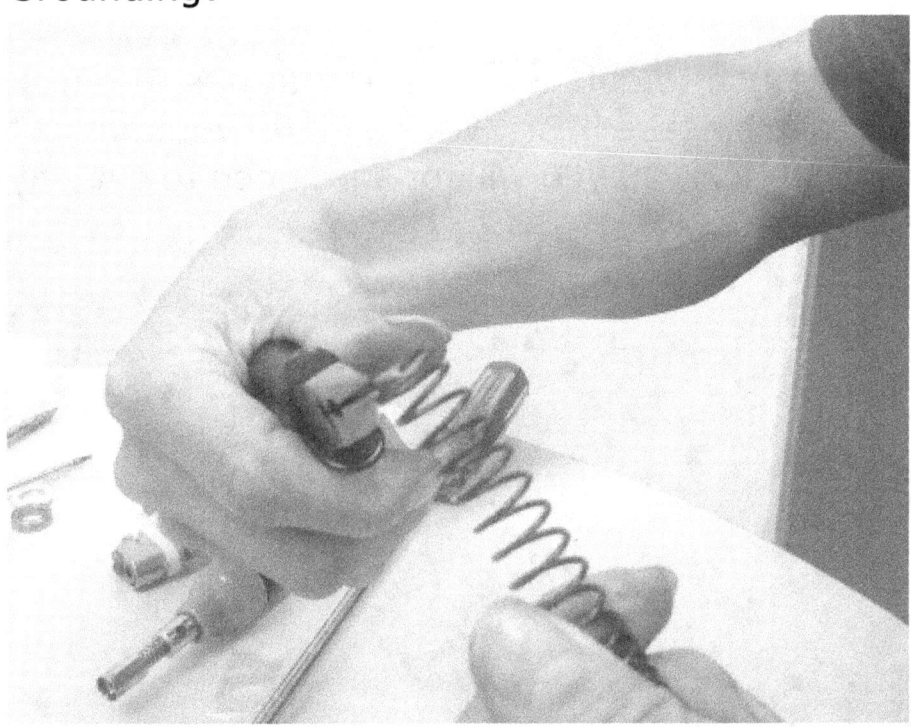

Smoothing and flattening:

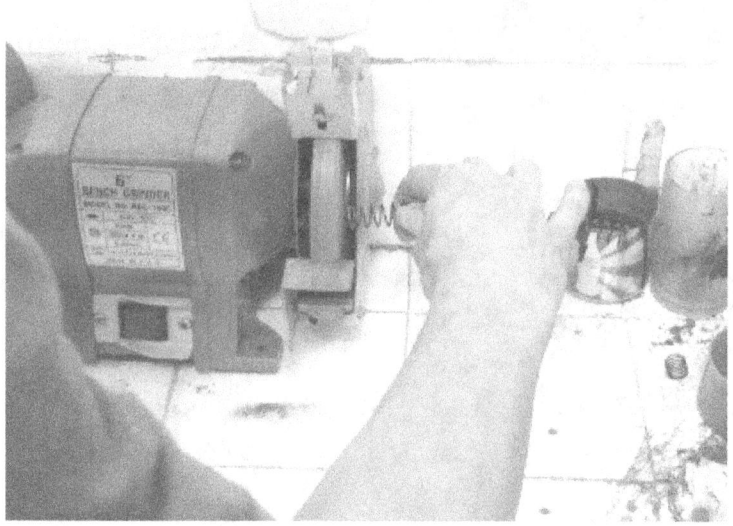

Do note that "how much to cut" is an issue that has to be determined on a case by case basis. With a M100 you may cut an inch off, while with a M120 you will for sure need to cut way more than that.

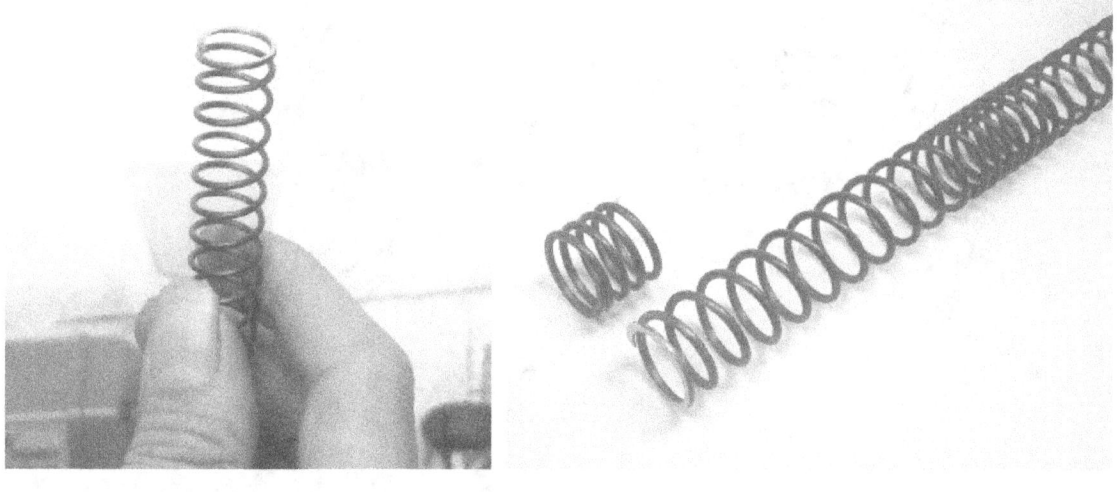

MechboxPRO AirsoftPRESS (Hong Kong).

Motor Tuning

You want to check the brushes. If the brushes are seriously worn out, replace them.

Different motors have different operating characteristics, that with different power source they may produce different RPM and different toque at different loads. And the efficiency does not have to be linear (refer to the example below):

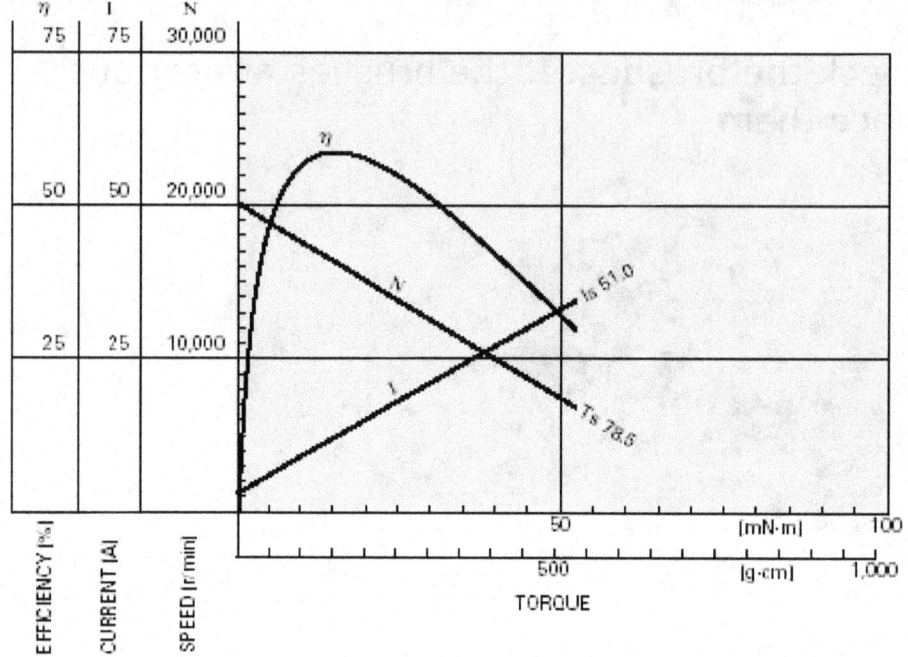

We do not want to go into details on motor technology. What we want to say is that, visual difference is minimal between most mass market motor models currently being offered in the market. If you want to see significant difference in torque or ROF, get a real expensive motor or change the gear ratio.

Do note that proper motor break in can improve overall performance. Refer to the special technote at the end of this book for further information.

Completing the re-assembly process

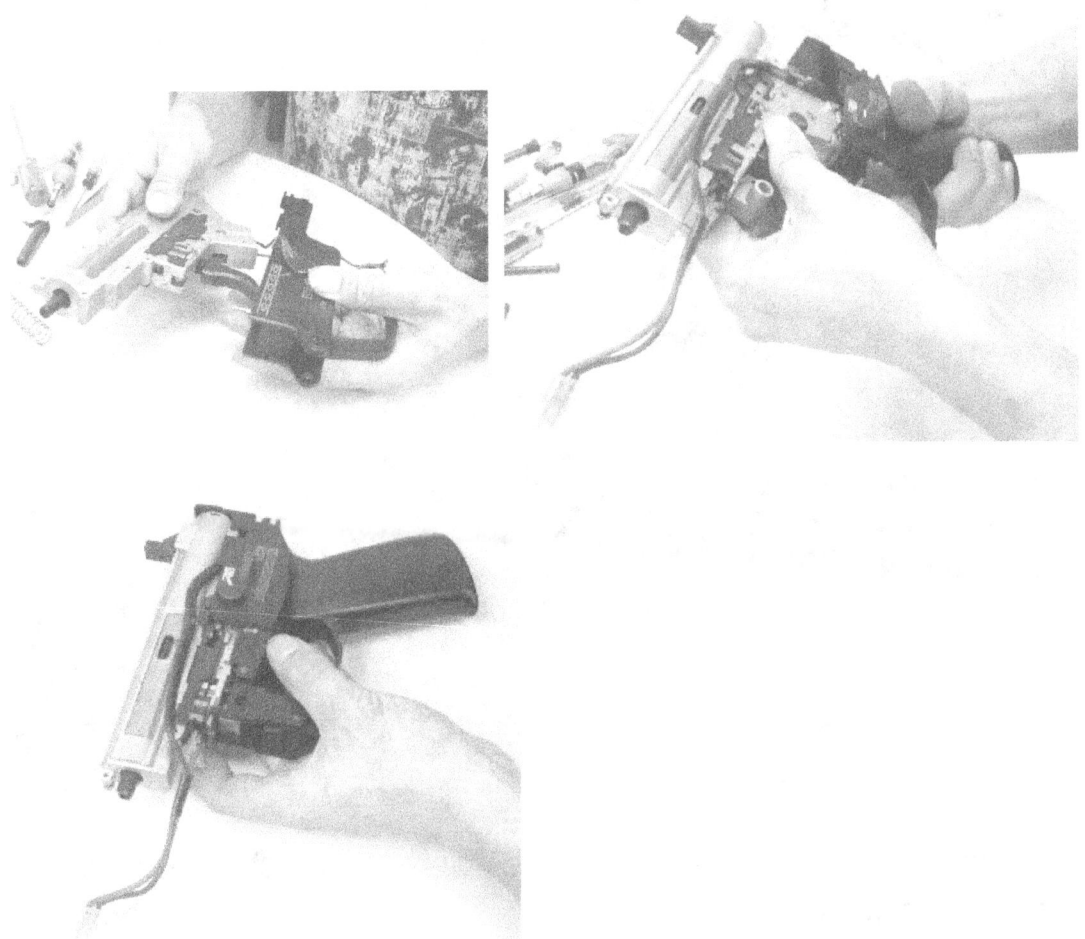

By going through the steps we described, a reasonable 1J configuration can be built.

Now let's move on to some advanced topics.

Using Sector Gear Delayer

The delayer must be installed in the correct direction – consult the manual that comes with it.

Keep in mind, tappet plate breakage could be possible when you are running a high ROF setup together with the use of a sector gear delayer. You can't do much in this regard – without a delayer you MAY misfire all the time (especially when the mag spring is not strong). Just make sure you buy a good quality delayer that will less likely produce troubles.

Not all delayers can fit into your sector gear. Always test fit them beforehand – a loose delayer can screw up the tappet plate instantly.

Note the tail of the tappet plate. If the shape does not interact smoothly with the delayer then a little bit of filing would be necessary. This is especially important if you plan to use a delayer.

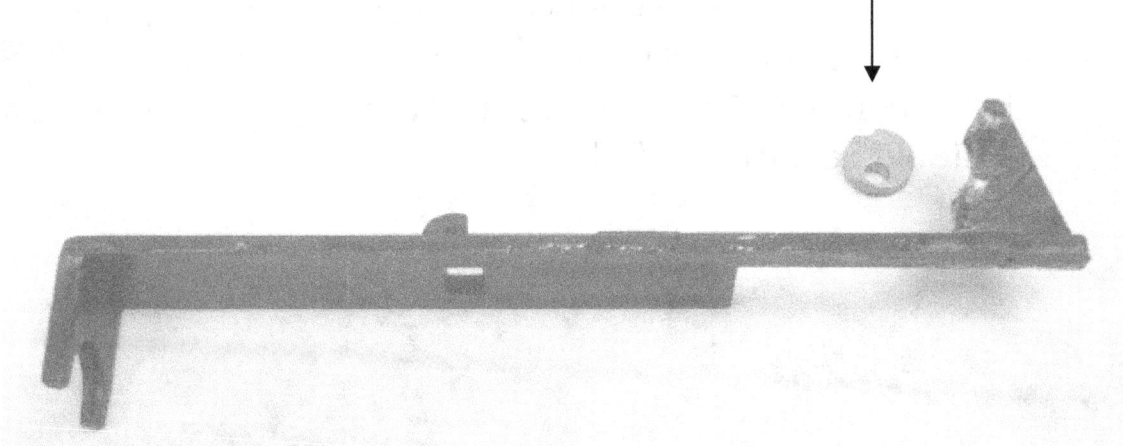

MechboxPRO AirsoftPRESS (Hong Kong).

Replacing the Switch Contact Plate

The switch contact plate can burn out after prolonged use and therefore would have to be regularly checked and replaced for maintaining current flow efficiency. If single shot is to be frequently exercised, you may want to consider replacing the switch plate as well. The plastic switch plate when under high heat can deform, which will eventually produce problem on the electrical contact. Switch plate made of high quality nylon plastic is highly recommended.

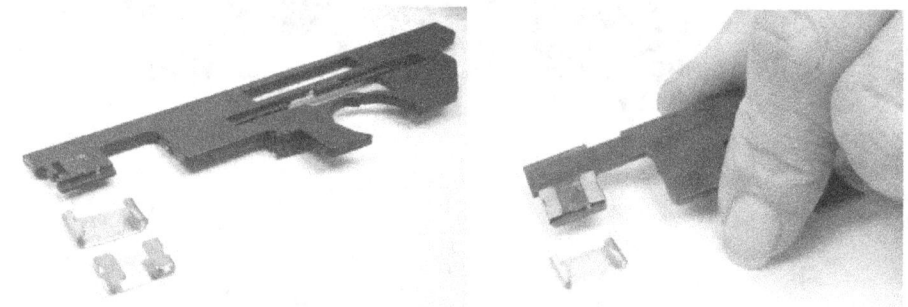

Good quality switch unit and connectors can reduce resistance and allow for efficient current pass through.

Single Shot Modification

For countries that restrict airsoft to single shot only, the modification must be reasonably "permanent" in a sense that the single shot mode cannot be overridden on the fly.

We recommend modification directly on the mechbox. A quick and easy way is to limit the switch plate movement:

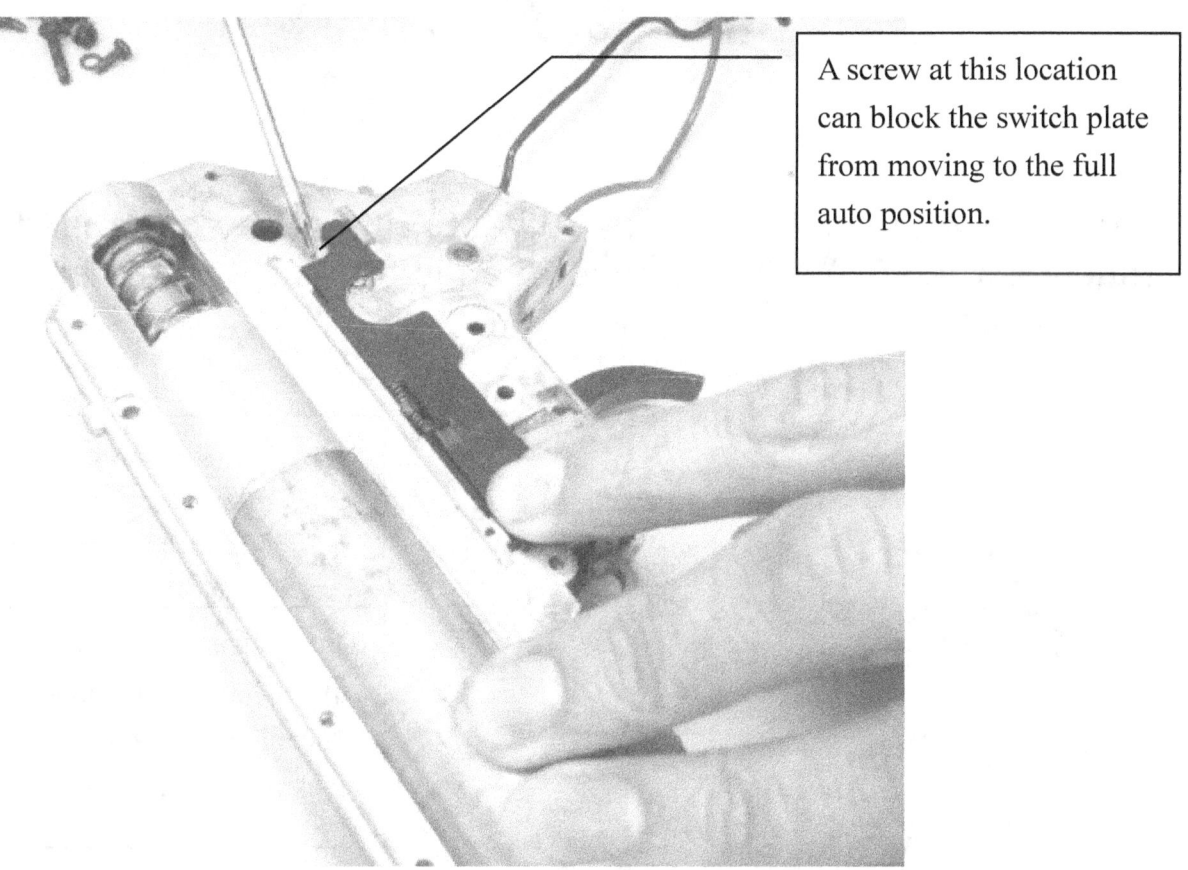

A screw at this location can block the switch plate from moving to the full auto position.

A more fancy and permanent way is to cut away some part of the switch plate so further movement cannot be possible:

MechboxPRO AirsoftPRESS (Hong Kong).

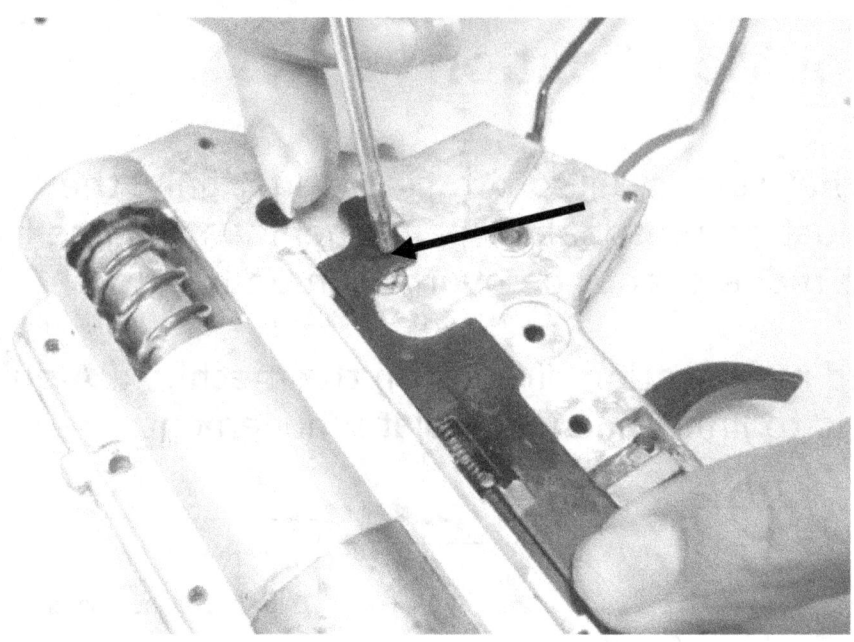

For those with single shot and full auto differently controlled (such as the P90), all you need to do is to block the full auto connection:

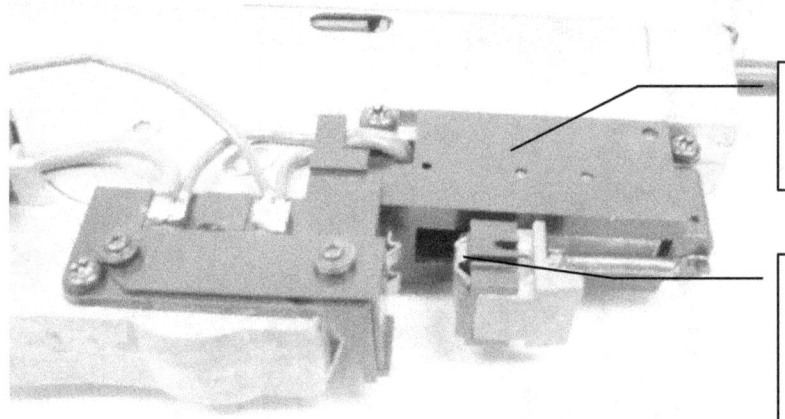

The hidden set inside is for semi.

This visible set is for full auto. Apply a layer of hot glue on either side of the connection can block full auto entirely.

The good thing about this approach is that it can be easily "undone".

Using High Speed Gears

Gears that are designed to mesh with each other usually come with identical tooth profiles. Through altering the standard ratio of teeth between the various gears, one may either enhance the final torque output or the final output speed.

There are different ratios among a pair of gears, and all we care is the final ratio of everything added up together (since we have 3 gears plus a motor pinion in the equation). Gearset with a lower final ratio can attain an increased rate of fire at the expense of final torque output, and vice versa.

Standard gear ratio is around (not exactly) 18:1. The high speed gears we use for demonstration is 13:1, with flat teeth. The spring that we use with them is a M100. The motor has a RPM of approximately 29000 under no load. Battery is the regular 9.6V 1100mah small cells pack.

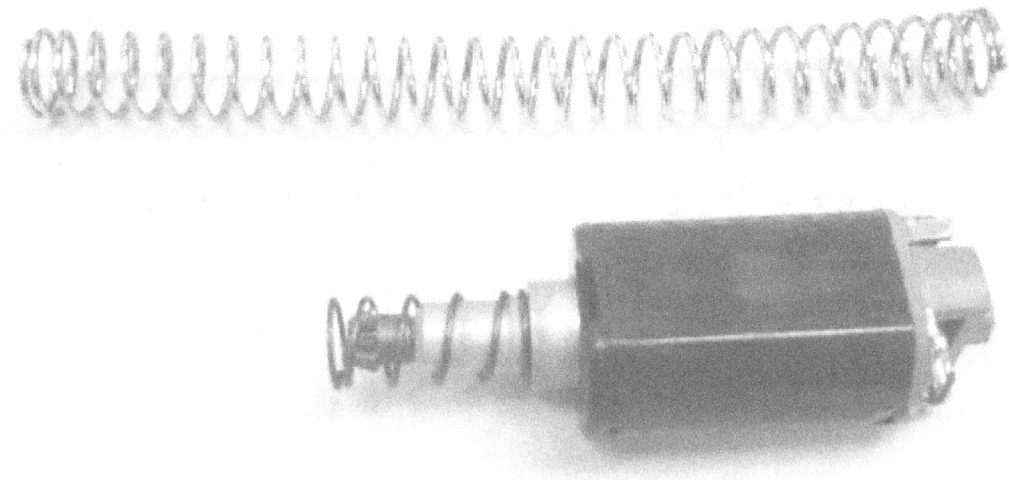

MechboxPRO AirsoftPRESS (Hong Kong).

Mixing and matching gears of different makes is NOT recommended due to likely difference in teeth shape:

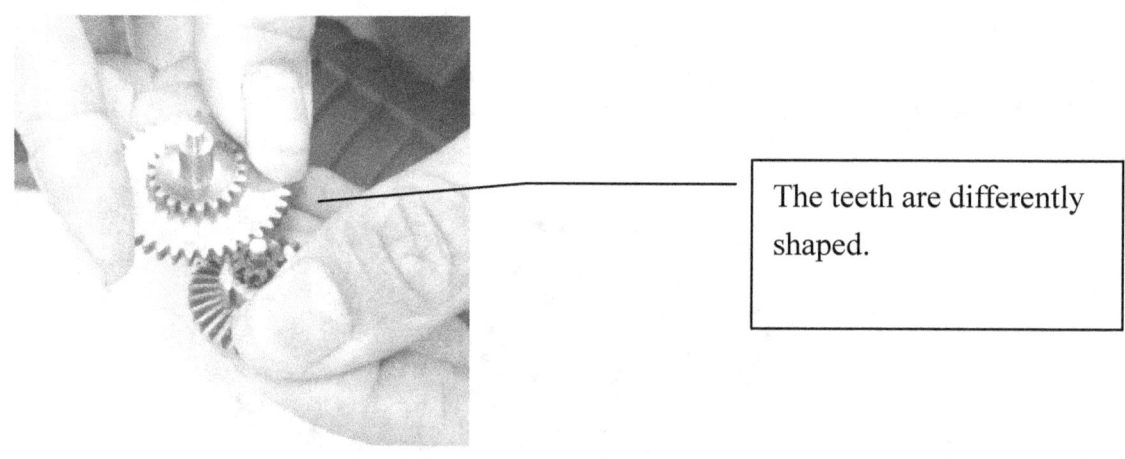

The teeth are differently shaped.

Below shows the comparisons between the standard gears (on the right) and the high speed gears (on the left):

Shimming these gears is no different from shimming the standard gears. First the spur gear, then the sector gear, and finally the bevel gear.

Do note that these gears are under more stress and will work under higher heat so there MUST be room left for flexibility.

Using high speed gears means there is less torque, so the motor will have heavier workload. You should therefore NOT use a high speed motor with these gears. If you do, you will need to push the motor real hard using higher battery voltage. Also, avoid stiff spring. Since the power limit we talk about is 1J, even a M100 is way too much.

MechboxPRO AirsoftPRESS (Hong Kong).

Further Troubleshooting

If after the upgrade your AEG shoots like this:

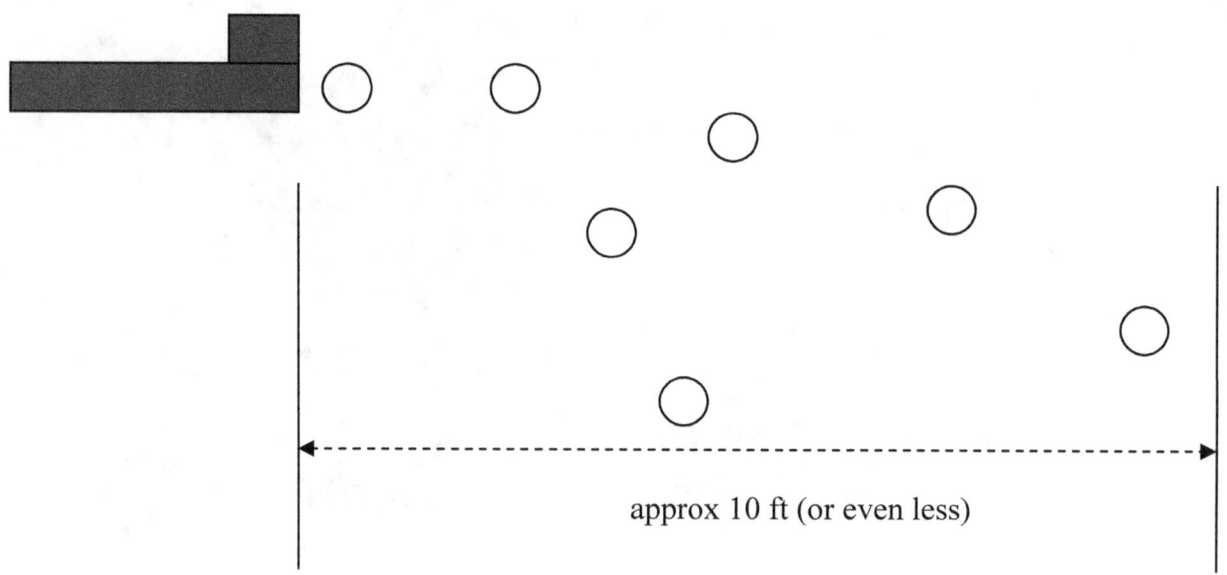

approx 10 ft (or even less)

<u>There are four major possibilities:</u>
i, damaged hopup - the rubber is dead, or the tiny spring which sustains the bucking is damaged. Practically speaking this seldom happens UNLESS you are using a China-made hopup rubber. There are many different grades of rubber and those China-made ones are just no good.

ii, grease is everywhere - at the tip of the air nozzle, around the hopup rubber, inside the barrel ...etc. **Highly likely after an upgrade job.** This is very easy to fix - just use the cleaning rod.

iii, extremely poor alignment between the air nozzle and the hopup chamber - this can happen if the mechbox has not been seated properly on the receiver (some wires get in the way). This has happened with many AK implementations. You just have to rewire carefully and reseat the mechbox.

iv, the air nozzle is too short. Replace the nozzle and retry.

If you experience instant gun lockup:

First use a high power 9.6V or 10.8V pack to "unlock" it. If it does not work and you see smoke coming out of the wires, something is wrong with the gears. Possibilities include:

i, a loose sector gear delayer. Some sector gears just don't work with the delayer, gluing it is of no use, so your only option is to remove it.

ii, over shimming at the bevel gear.

iii, you left a screw somewhere inside the mechbox.

iv, a tappet plate that does not fit well with the shell. You either sand the sides of the tappet plate to make it fit, or replace it with one from another manufacturer.

If you experience gradual gun lockup, possibilities may include:

i, a loose sector gear delayer. Remove it.

ii, over shimming somewhere (or everywhere).

iii, poor quality / aging wires and/or connector plugs (not likely the case if your gun is pretty new).

iv, a spring that has not been properly grounded on either end (or on both ends). Refer to the picture below, the tip of the spring can prevent the spring itself from spinning, which could produce unnecessary tension on the gears. Check and make sure both ends of the spring are grounded (they should be flat), and that there is at least one metal ring spacer on the stock plastic spring guide to allow smoother spinning (apply a little bit of grease on both ends of the spacer too).

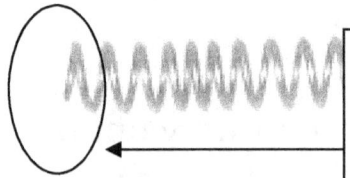

You may ground it yourself. Just clamp the tip and the last coil together and then heat it up with a lighter for 15 seconds or so.

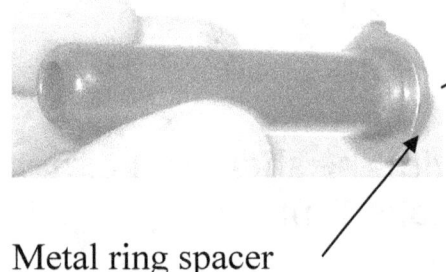

You don't need to use any spacer if you have a ball bearing spring guide.

Metal ring spacer

At the inside of the piston, if there is a spacer then make sure the surface of the spacer is completely flat. If you are using the TM type spacer (see the picture below), file the surface to remove any unnecessary debris on it.

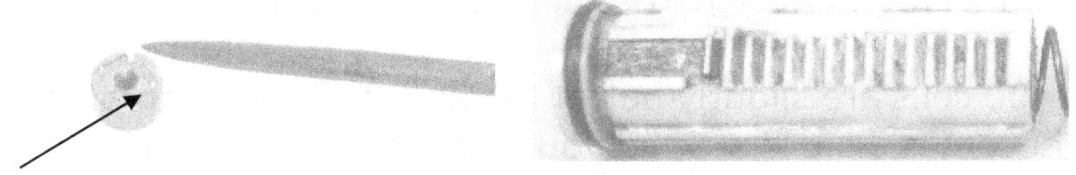

If the first 2 teeth of the piston break quickly:

i, your battery is way too powerful for the setup

ii, your piston is too brisky (this often happens with the China-made pistons).

Keep in mind, you won't fry your mechbox by using high power battery. You will only fry your mechbox ("burn" the switch plate) if something gets stuck inside.

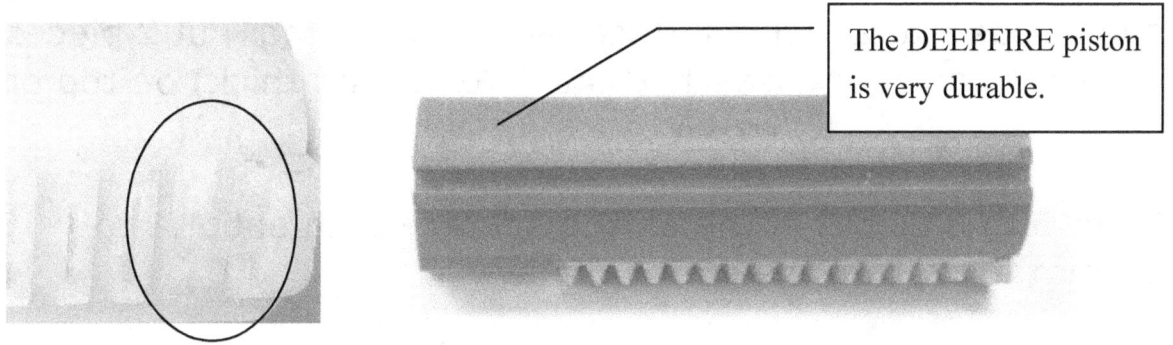

The DEEPFIRE piston is very durable.

What if you have no choice other than the small 1100mah cells?

Believe me, with these small cells even a 12V pack is very marginal for a PDI 170% (voltage will drop very quickly and the battery will get heated up like hell). This is especially true when your cells are from an unknown brand – many China-made battery clones are of very poor quality and would give you a way-faster-than-expected voltage drop.

The 1400 cells are slightly better but still, due to the small size heat becomes a big issue and your battery pack won't last long for even a half day game. Solution? A dirt cheap solution (assuming you already have the right air seal) is to do ALL of the following:

i, get a piece of stock spring from a WELLs/CYMA AEG. This stock spring is softer than a PDI spring, but is still stronger than a stock TM spring when configured with proper compression (through the use of spacers). The spring is easier to pull and is easier for the entire power system, especially when your battery is not in peak condition.

ii, use the stock TM piston, retain the piston spacer but make sure it has a totally flat surface facing the spring. Put 2 pieces of metal ring spacer (each about 1 to 1.5mm thick) on top of it.

iii, fit 5 to 6 pieces of metal ring spacer (each about 1 to 1.5mm thick) into the spring guide

Install the above into your mechbox and the gun will be able to produce the expected power without sacrificing reliability.

Special Topic: Maximum Air Seal

The importance of air sealing and air tightness

The goal of proper air sealing is to fill every gap along the air flow path to minimize air leakage and create an efficient environment for propelling BBs out of the barrel. To achieve this goal you need to first test for air tightness. Based on our experience higher end products from Marui, CA, ICS and the like are usually pretty good at that, but you can't expect the same from those cheaper China-made AEGs. Air sealing shall therefore become your top priority in gun upgrade and tuning.

NOTE: With proper air sealing you can easily achieve 10~15% of FPS improvement on your China-made AEGs almost instantly.

Air tightness is critical in the following two areas of the air flow path:

- Cylinder set
- Barrel and Hopup

Air tightness at the cylinder set

To test for air tightness at the cylinder set, block the cylinder head nozzle with a finger and then try to push the piston.

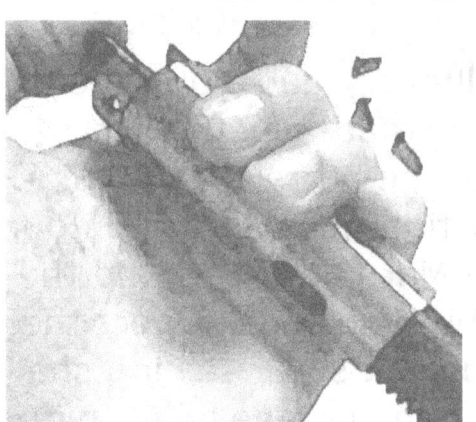 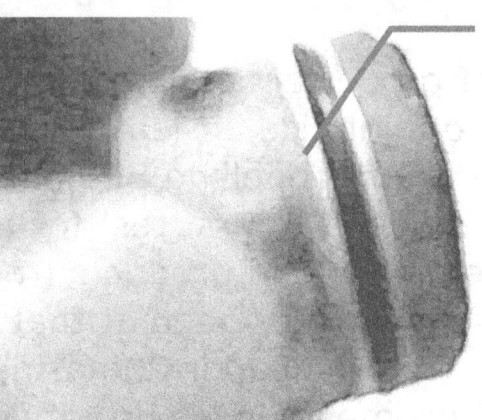

Pipe wrap behind the O ring

If you experience strong resistance, air tightness is good. If not, two possibilities exist:

I, Air leaks through the gap between the cylinder head and the cylinder. To prevent leakage of this sort, you use a thin layer of pipe wrap to wrap around the side of the cylinder head before putting in the cylinder head O ring:

II, Air leaks through the piston O ring (this is the most common cause of performance problem among the China-made AEGs). A properly sized replacement can usually fix the problem.

Air tightness at the barrel & chamber set

To test for air tightness at the barrel and the hopup chamber, you block the chamber entrance and the BB loading nozzle with your fingers and then try to blow air into the barrel:

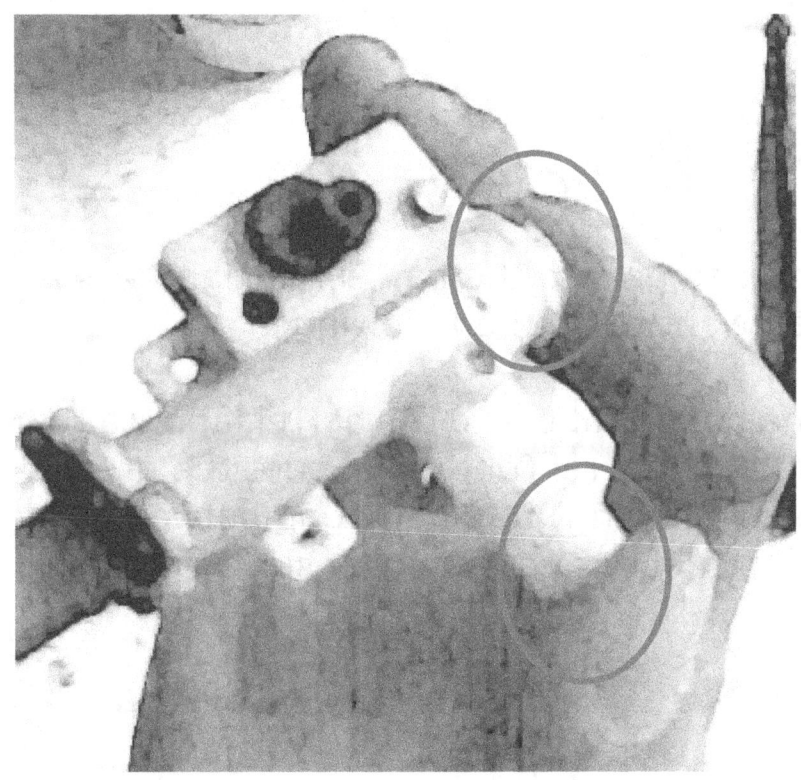

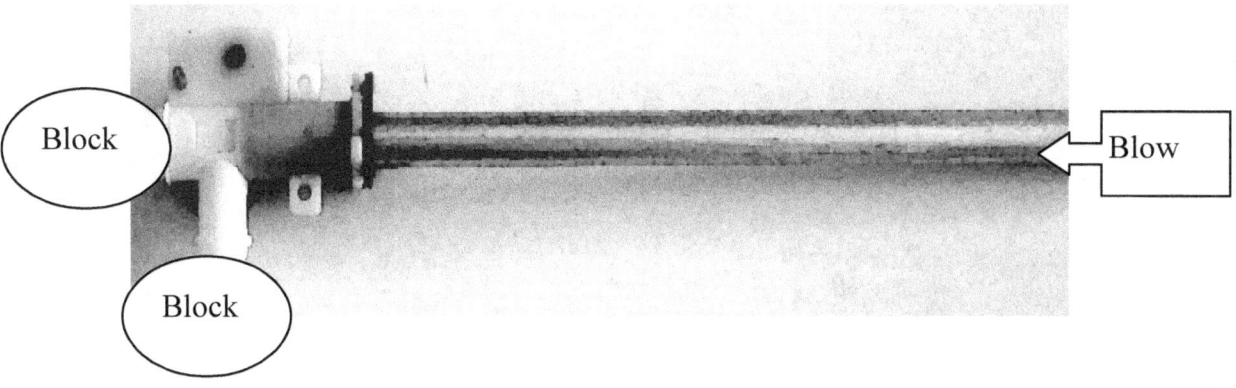

Block

Blow

Block

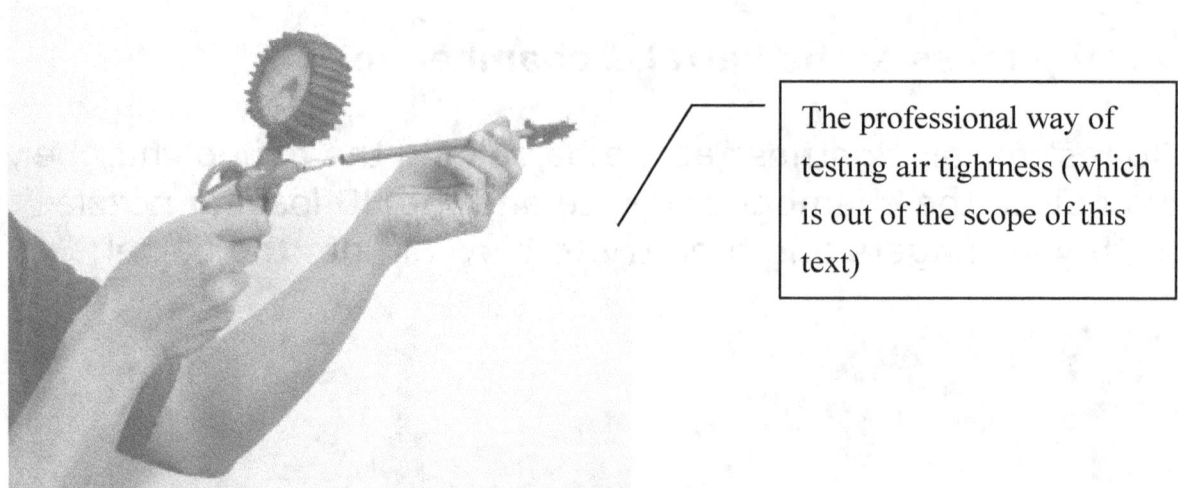

The professional way of testing air tightness (which is out of the scope of this text)

If it is real tough for air to go into the barrel, air tightness is good. If not, you may want to do the following:

I, Before attaching the hopup rubber to the barrel, apply a thin layer of grease around the barrel (but make sure the grease doesn't block the barrel opening) to fill any gap between the inner side of the hopup rubber and the outer surface of the barrel:

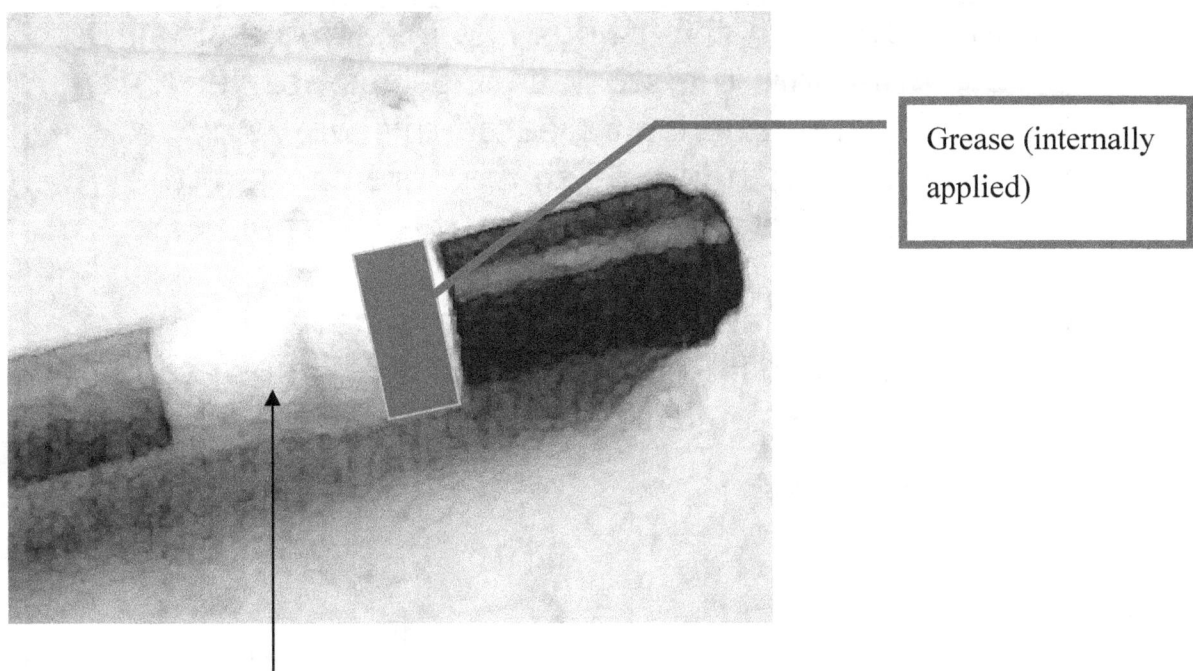

Grease (internally applied)

II, Now use pipe wrap to wrap around the barrel and the hopup rubber to further prevent air leakage.

III, You may also want to improve air tightness at the air nozzle. The easiest thing to do with your stock nozzle is to apply a very thin layer of grease around the inner side of the nozzle (and most importantly - without blocking the opening):

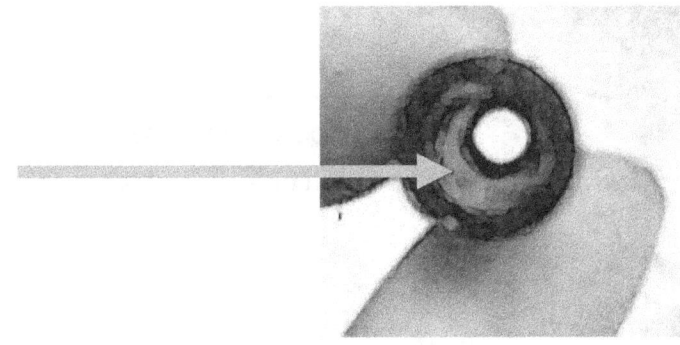

Warning: Don't do silly things like inserting a small O ring into the stock O-ring-less nozzle. An air seal nozzle isn't created this way. Any mismatch in size/shape here can jam the mechbox and screw up the internals.

High quality nozzles always have O rings "built-in"! Refer to the picture below, the left one has built-in O ring. The right one does not.

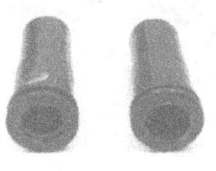

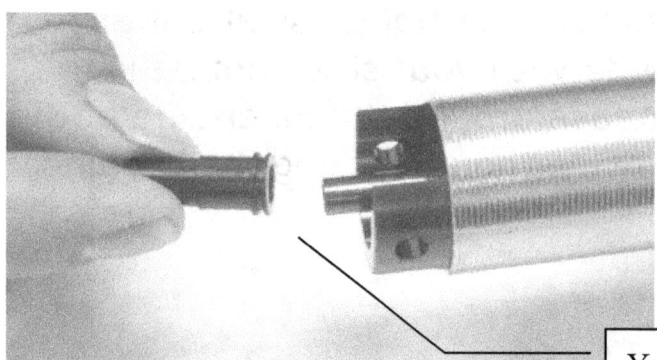

You must be very precise here – air leakage can be very serious at the nozzle.

MechboxPRO AirsoftPRESS (Hong Kong).

Special technote on piston head

modification

Upgraded piston heads usually have holes on the front face of the head. When the piston is moving, air goes through these holes and forces the surrounding O-ring outwards a tiny bit to create a seal for preventing air leakage (this is how the one-way mechanism works).

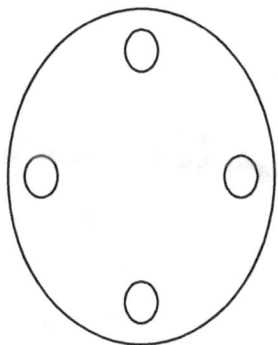

The WELLS piston head has the holes drilled for you already so you don't need to modify the piston head any further (except to replace the O ring when necessary). The APS one even has 6 holes on it. Frankly, more holes on it would not be necessary. On the other hand, make sure you don't over-apply grease or these holes will be accidentally blocked.

The Pro Ology high end piston head has a different design – it has a proprietary one-way design where the holes are actually hidden behind the O ring.

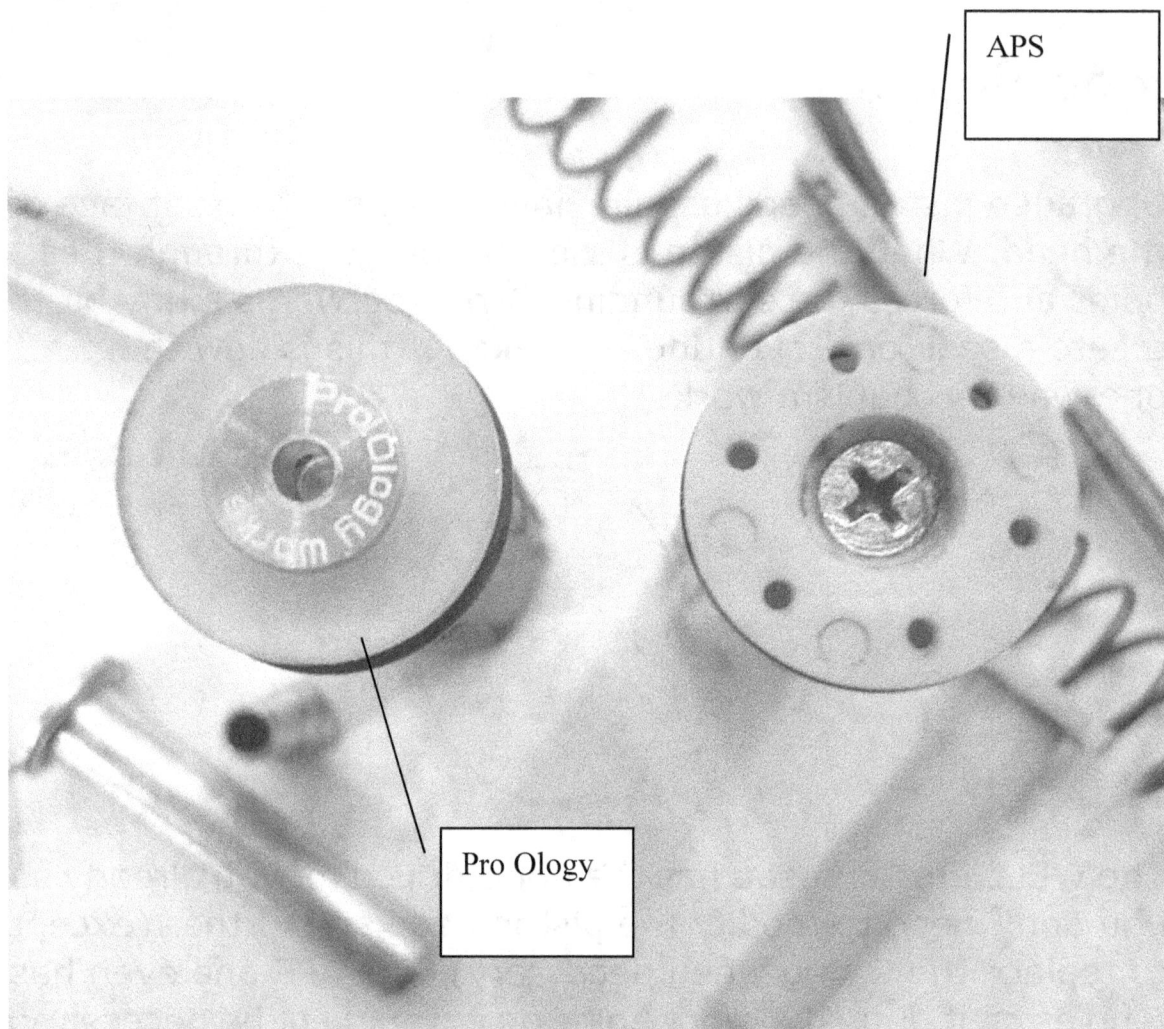

APS

Pro Ology

Special technote on air nozzle shape and sizing

Improper nozzle configuration may lead to serious air leakage or even hopup breakage. The pictures below show 3 different problems you may encounter when configuring your AEG (full credit goes to zhigangd2005 who posts these valuable pictures at http://forum.combat2000shop.com/:

Problem 1: Somehow the hopup chamber and the nozzle are not leveled (most likely the hopup chamber has not been properly installed, OR the tappet plate was bent, resulting in damaged hopup rubber.

Notice the angle of the nozzle

Problem 2: The nozzle is too short, resulting in air leakage.

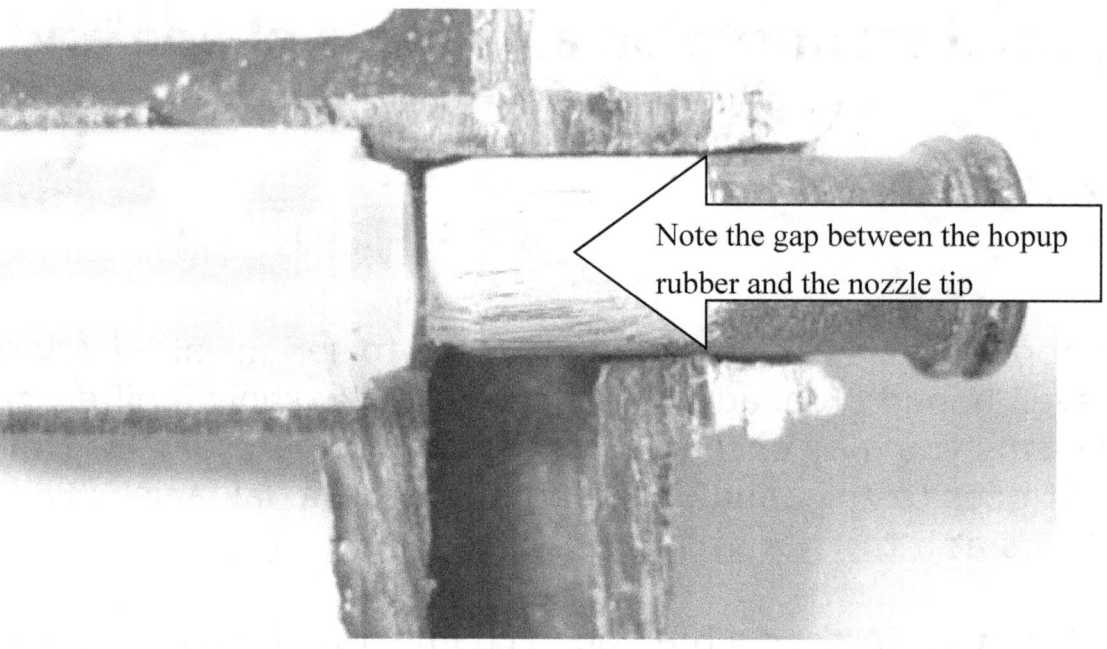

Note the gap between the hopup rubber and the nozzle tip

Problem 3: The air nozzle is too long, and the nozzle tip has a shape that does not fit well with the opening of the hopup rubber.

Mismatch between the nozzle tip and the hopup rubber

The picture below shows 3 different air nozzles produced by 3 different Chinese manufacturers for the same gun model. They are all different, even though their hopup chambers are similarly structured.

The **DEEPFIRE M4** implementation is among the BEST in the industry. Its standard hopup unit and the enhanced one-piece unit can provide optimal air seal out of the box.

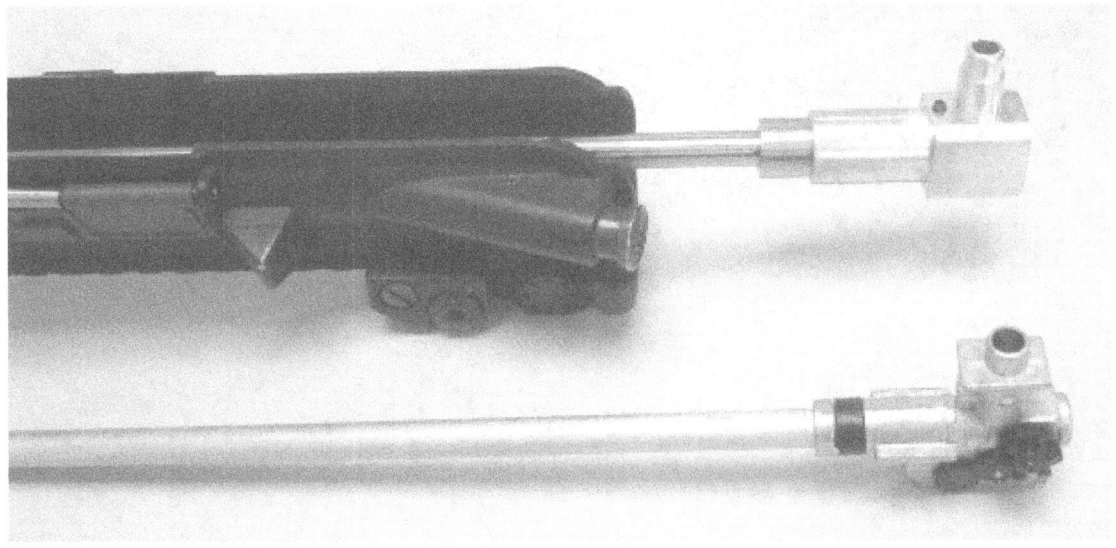

MechboxPRO AirsoftPRESS (Hong Kong).

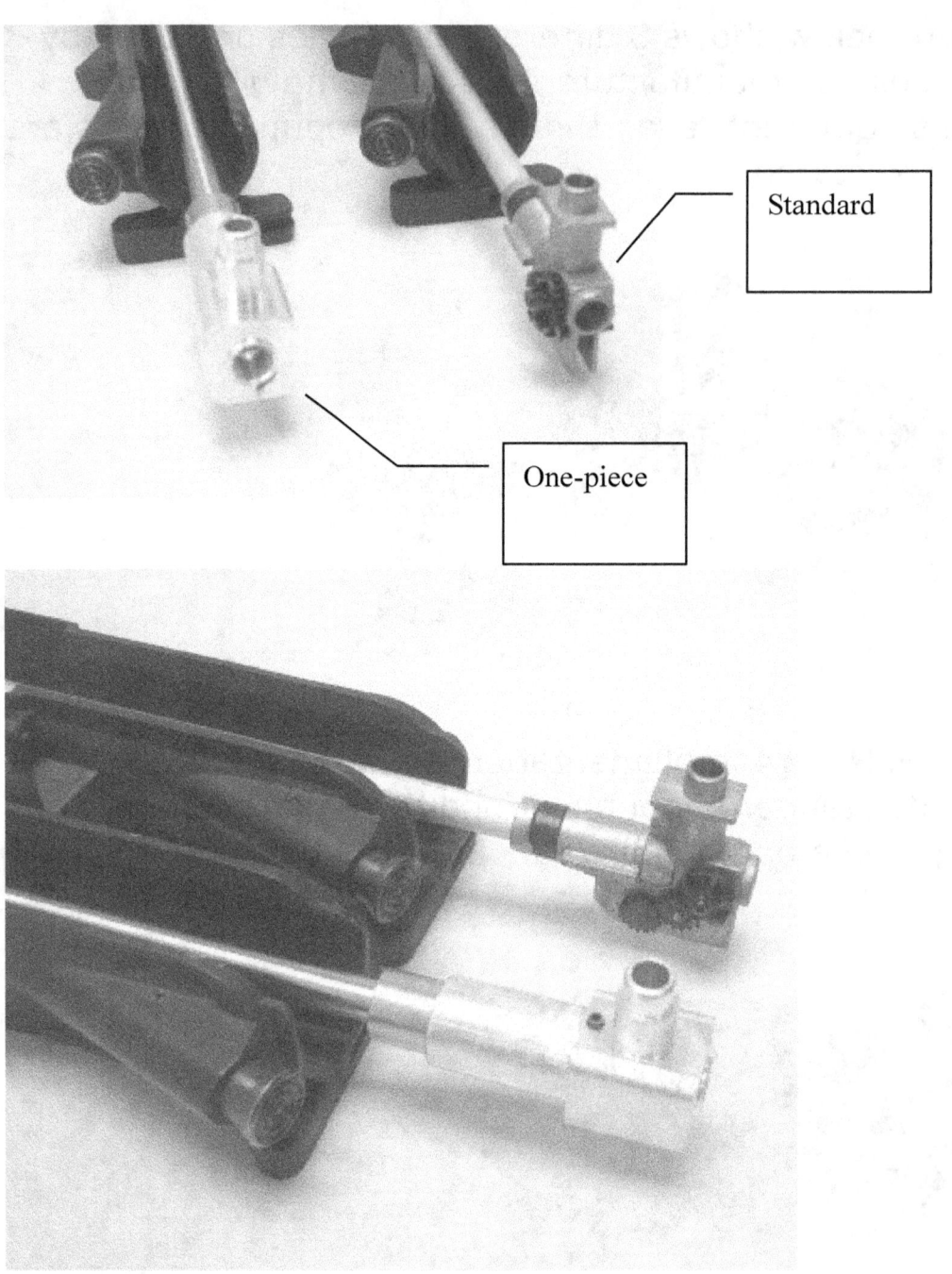

Standard

One-piece

MechboxPRO AirsoftPRESS (Hong Kong).

Special technote on motor and gears

break-in

You want to first break-in the brush/commutator interface so that the brushes can conform better to the shape of the commutator. To do so is easy – just run the motor on 4 cells (1.2V on each cell) for several minutes until the full brush face is conformed to the commutator.

Breaking in bushings is necessary if the motor bushings are too tight. To do a quick check, just spin the motor (by hand) with the brushes removed and fell the resistance to turning (you may want to have some other motors here for comparison purpose).

To perform bushing break-in, just put a little valve of grinding compound into the bushing and spin the motor until you feel a reduction in resistance.

MechboxPRO AirsoftPRESS (Hong Kong).

You may also want to break-in the mechbox gears for smoother operation. To break in these gears, setup the mechbox in such a way that only the bushings and the gears are in place. No anti-reversal, no piston and no other parts. Then, close the mechbox shell tightly and hold the motor to the bevel gear. Have the motor driven by 6V or 7.2V power (with excellent shimming your motor should be able to drive the gears at 6V), and let the motor pinion drives the gears. Of course, before you do this you want to ensure that all these gears (including the motor pinion) have been lubed. Let them spin like this for a couple of minutes and the gears break-in process is considered completed.

MechboxPRO AirsoftPRESS (Hong Kong).

For the latest product releases, please visit:

http://www.airsoftpress.com

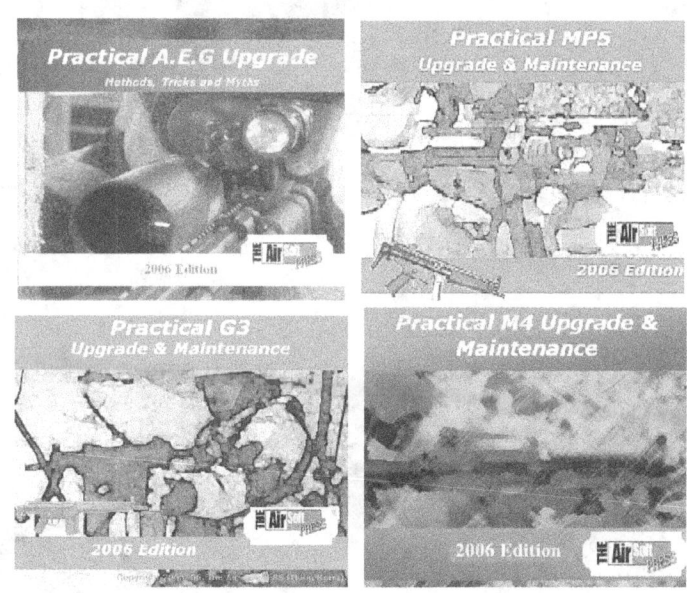

Please email your questions and comments to

editor@airsoftpress.com

Thank you.

DEEPFIRE offers perfect quality V2/V3 mechbox components. It also has a line of entry level M4 AEGs (the Silver series) with exactly 1J power.

RCPRESS publishes books on RC technology.

www.rcpress.com

G-Military aimed to provide the greatest online purchasing experience for fans who love gun and wargame. Its AIP line of AEG motors offers high performance at very affordable prices for the V2 mechboxes.

In terms of bearings and bushings, their 7.03mm steel bushings are perfect for most China-made mechboxes.

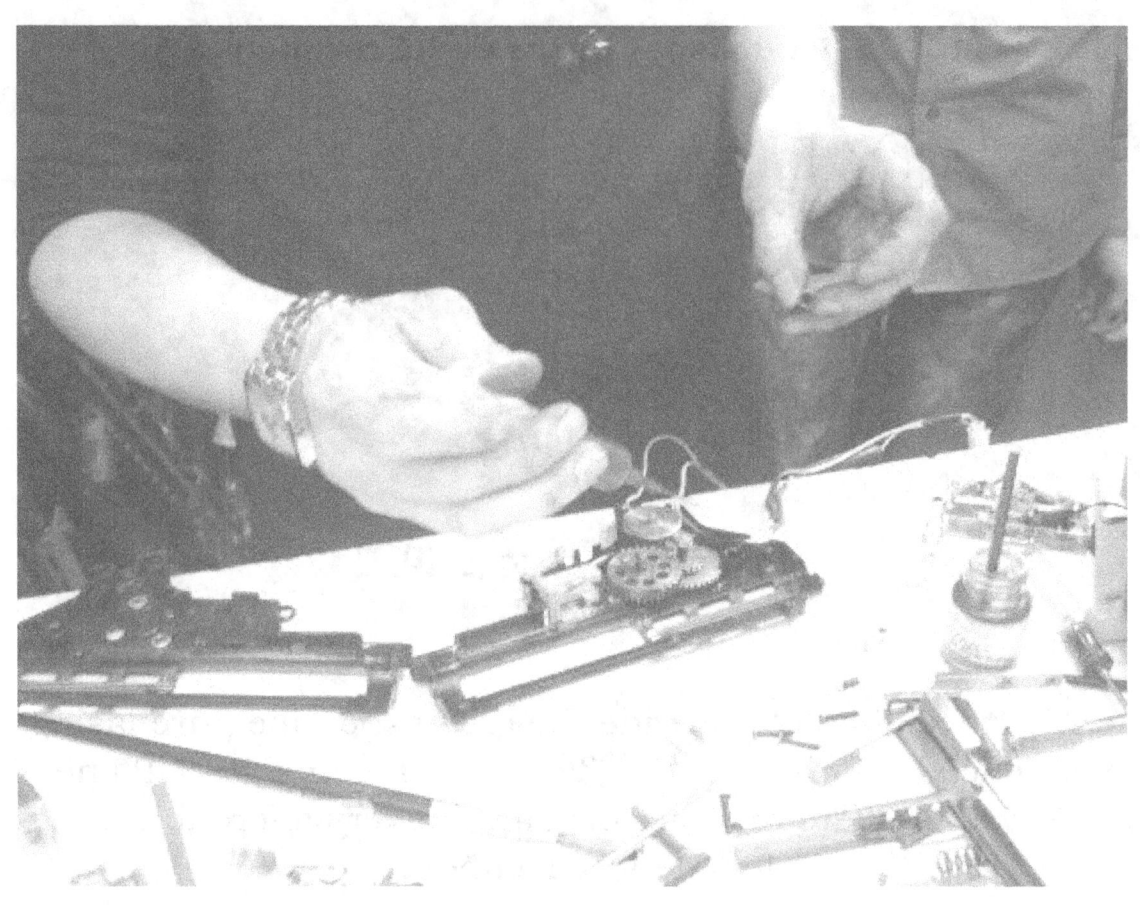

Pro.Ology Works
Engineer Republic

Pro.Ology Works offers first class upgrade service on airsoft products.

www.ingramcontent.com/pod-product-compliance
Lightning Source LLC
Chambersburg PA
CBHW081213170526
45165CB00009B/2805